Strategies for Acquisition Agility

Approaches for Speeding Delivery of Defense
Capabilities

PHILIP S. ANTON, BRYNN TANNEHILL, JAKE MCKEON,
BENJAMIN GOIRIGOLZARRI, MAYNARD A. HOLLIDAY,
MARK A. LORELL, OBAID YOUNOSSI

Prepared for the United States Air Force
Approved for public release; distribution unlimited.

 PROJECT AIR FORCE

For more information on this publication, visit www.rand.org/t/RR4193

Library of Congress Cataloging-in-Publication Data is available for this publication.
ISBN: 978-1-9774-0436-7

www.rand.org

Preface

With increased capabilities by potential adversaries and advances in technology, the Department of the Air Force (DAF) and the rest of the U.S. Department of Defense, along with Congress, have pushed for ways to more rapidly acquire capabilities in general and space systems in particular. Anthony Reardon, then–Assistant Deputy Chief of Staff for Strategic Plans and Requirements in AF A5/8, asked RAND Project AIR FORCE to analyze approaches for improving acquisition agility with an emphasis on space acquisition. Rather than trying to invent a silver bullet for speeding up acquisition, we sought to review and assess the variety of more-responsive and more-agile approaches that have proven themselves over the decades or are emerging as new experimental paradigms. This report presents a framework for selecting approaches relevant to the characteristics of the acquisition to be conducted. In addition to creating this integrated view, the project integrated lessons learned on whether the DAF recompetes prime system integrators for systems with the same essential architecture. Research for this project was conducted from October 2017 through November 2018.

The research reported here was commissioned by AF A5/8, Headquarters Air Force, Washington, D.C., and conducted within the Resource Management Program of RAND Project AIR FORCE as part of a fiscal year 2018 project titled *Agile Acquisition processes*. This report should be of interest to those involved in efforts to speed the acquisition of space systems and other defense capabilities.

RAND Project AIR FORCE

RAND Project AIR FORCE (PAF), a division of the RAND Corporation, is the U.S. Air Force's federally funded research and development center for studies and analyses. PAF provides the Air Force with independent analyses of policy alternatives affecting the development, employment, combat readiness, and support of current and future air, space, and cyber forces. Research is conducted in four programs: Strategy and Doctrine; Force Modernization and Employment; Manpower, Personnel, and Training; and Resource Management. The research reported here was prepared under contract FA7014-16-D-1000.

Additional information about PAF is available on our website: http://www.rand.org/paf/.

This report documents work originally shared with the U.S. Air Force on December 7, 2018. The draft report, issued on December 21, 2018, was reviewed by formal peer reviewers and U.S. Air Force subject-matter experts.

Executive Summary

The U.S. Air Force and U.S. Department of Defense have employed a variety of approaches for more-responsive acquisition of military capabilities given evolving threats and technology opportunities. The Assistant Deputy Chief of Staff for Strategic Plans and Requirements in AF/A5/8 asked RAND Project AIR FORCE (PAF) to identify and analyze various approaches, assess their suitability for different conditions and types of acquisition, and identify implementation issues.

Issues

- What approaches can be employed to speed acquisition or minimize schedule growth on acquisition programs?
- How can program managers and stakeholders identify the best approach (or approaches) for a given type of acquisition?
- What practical matters arise during implementation of these approaches?

Study Methodology

Rather than try to invent a silver bullet, PAF researchers assessed the variety of agility approaches that have proven effective over the decades or are emerging as new experimental paradigms. The team developed a selection framework and tool that helps program managers identify relevant approaches.

Conclusions

- Few agility approaches are universally applicable. The right one for a given acquisition depends on the conditions for application, the domains involved (e.g., requirements, budgeting, acquisition), and issues with implementation.
- Many approaches are tried and true and require no special authorities to implement.
- Agility depends not just on acquisition but also on requirements, budgeting, technology, and intelligence activities.
- Speed could still involve compromise in cost or technical performance objectives.

Opportunities

- Program managers and stakeholders can use the PAF-developed spreadsheet tool (see figure) to identify relevant agile approaches when developing acquisition strategies or structuring organizations.
- Investment in workforce expertise and experience, ready availability of financial resources, and a willingness to accept operational capabilities incrementally are important factors in agility, in addition to process improvements.

The spreadsheet tool looks up which acceleration techniques (out of 63 considered) might be appropriate for the conditions selected (out of 49 possible, plus those that are universally applicable). In this notional example, the user has selected the six conditions in green, and the tool suggests the three approaches in yellow as potentially applicable.

Acquisition Acceleration Approaches Lookup Tool

Clear All Selections

Select your conditions:

NONE (universal)	Needs bidding approach beyond FAR
Able to influence budgets	Needs strategic partnership beyond FAR
Able to utilize existing capabilities	Operators can tolerate unknown risks of system failure
Budget needs relatively small in near term	OTA (Other Transaction Authority) available
Budgetary resources obtainable	Production can be quick
Commercial capability or technology of interest	Program small dollar value
Commonly needed contractor good/service	Prototype successful
Contracting areas known for future	Prototyping possible in short time
Contractor PM exceptionally experienced and skilled	Requirements approval authority at the Component level
Contractor prequalification possible/needed	Requirements: flexible level of acceptance
Contractor team highly skilled	Requirements may be tradable given cost/schedule/issues
Empowered decision makers	Requirements priority high
Explore CONOPS	Requirement is urgent or emerging (e.g., (J)UON/(J)EON)
FAR exemptions consistently applicable	Risks low (generally)
Formal feedback loop for training or education	Risks of technology high
Funding available in prototyping account	Security risks from foreign content is low
Government development	Software dominant
Government PM exceptionally experienced and skilled	Sole-Source (FAR 6.302) applies
Government staff capacity	Suboptimal, modular architecture and standards are sufficient
Government staff highly skilled	System modifications or alternative production contemplated
Incremental capabilities useful	Technical development needed and significant
Intellectual property protections beyond FAR	Technology relatively mature
Involve users in exploring capability options	Threats are changing
Joint issues largely absent	Value determination needed
Learning curve high or infrastructure costs high	Well-defined good/service

These agile approaches may be applicable for your conditions:

Skunk Works®-like organizations

"Crashing" the schedule

Senior board-of-directors (direct approval / oversight to PM)

NOTES: CONOPS = concept of operations; FAR = Federal Acquisition Regulation; (J)EON = (Joint) Emergent Operational Need; (J)UON = (Joint) Urgent Operational Need; PM = program manager.

Contents

Boxes, Figures, and Tables

Boxes

Figures

Tables

Summary

Long acquisition times have been a significant concern for the U.S. Department of Defense (DoD) for decades. It is critical to provide capabilities to warfighters in a timely manner relative to the threats faced, and various approaches have been taken to reduce acquisition timelines. To reduce the time required to field operational capabilities, various Department of the Air Force (DAF) and other DoD organizations have used a wide variety of approaches to acquisition that are more responsive and more agile. These organizational strategies for accelerated acquisition draw on multiple *approaches* (e.g., techniques, methods, processes, resources, and organizational constructs), including some that can reduce acquisition time compared with norms and others that can mitigate (minimize) schedule growth. In this report, we describe new and evolving approaches along with historical approaches and those in practice as of late 2018. This diversity, however, can make it difficult for program managers (PMs) and stakeholders to identify approaches that are appropriate for a given situation and type of acquisition, so we describe four criteria for selecting these candidates—but the ultimate choice depends on judgments by decisionmakers. Finally, we conclude with perspectives on the fundamentals of acquisition and how trade-offs, risks, and other issues factor into speeding acquisition.

Identifying and Selecting More-Responsive Acquisition Approaches

In our analysis of DAF and other DoD rapid-acquisition organizations, we found a wide variety of approaches for more-agile acquisition (Box S.1). Our analysis consisted of reviews of the literature, our own experiences, case studies of rapid-acquisition organizations, examination of approaches and challenges of interest to our sponsor, construction of a selection framework that maps application conditions to more-responsive acquisition approaches, and identification of implementation considerations for six areas (acquisition strategies, acquisition processes, program structures, headquarters structures, requirements, and budgeting), as well as downsides and trade-offs. The existence of these approaches (including many that are not new) together with the existence of many rapid-acquisition organizations illustrates that the acquisition community indeed knows how to go faster if speed is of the essence. Further analysis of these approaches identifies key conditions necessary for implementation.

In assessing these approaches and their properties, we found at least four different factors that can be taken into consideration when selecting which approach to use:

- the necessary **conditions** for application
- practical **implementation considerations** (i.e., what needs to be done in acquisition strategies, acquisition processes, program structures, headquarters structures, requirements and budgeting to support the approach)
- the **domains** involved (especially requirements, budgeting, and technology)
- the **stage** in the acquisition's life cycle at which the approach could be used.

Box S.1. Approaches for More-Responsive Acquisition, by Category

Higher-Level
- Acquisition tailoring
- Technical risk mitigation

Combined
- Skunk Works–like organizations
- "Agile" development

General
- In-house engineering and production
- Formal, active lessons learned process
- Standing FAR waivers

Intelligence-Related
- Faster intelligence

Requirements-Related
- Requirements: partial (80%) solutions acceptable to users
- Requirements: keep stable (avoid "creep")
- Configuration steering boards
- Bypass JCIDS and JROC

Schedule-Based
- Crashing the schedule
- Fast-tracking (parallelization; concurrency)
- Streamlining
- Focused work
- Work-breakdown schedule reviews
- Schedule reserve (margin; contingency)
- Readiness-based gates or milestones
- Hard, fixed schedules

Financial-Related
- DoD Rapid Prototyping Fund
- Reprogramming
- Budget stability measures
- Management reserves

Technology Development–Related
- Operational prototyping: new platforms
- Operational prototyping: middle tier of acquisition rapid prototyping (10 USC §2302 Historical and Revision Notes; Pub. L. 114-92, §804 [hereafter referred to as "Section 804"], as amended, 2015)
- Operational prototyping: components
- Operational prototyping: components or technology (10 USC §2447; Pub. L. 114-328, §806, 2016)
- Joint Concept Technology Demonstrations (JCTDs)
- Prize competitions

Reuse
- Modifications to existing platforms
- Repurposing
- Off-the-shelf: COTS or GOTS
- Adaptation of foreign technology

Maturity-Based Modification
- Incremental acquisition
- Continuous component upgrade as technology matures

Design-Related
- Modular open-system architectures
- Disaggregated architectures
- Own the technical data (especially interfaces)

Testing-Related
- Test to acceptance
- Operational testing in actual operations

Box S.1—Continued

Rapid-Fielding

- Rapid fielding: middle tier of acquisition (10 USC §2302 Historical and Revision Notes; Section 804)
- (J)UONs or (J)EONs

Contracting-Related

- Indefinite-delivery contracts (e.g., IDIQs)
- Long-term prime systems integrator
- Noncompetitive contracting: justification and approval
- GSA schedule
- Marketplaces
- Bids open to consortium members only
- Sole-source production following successful prototypes
- Custom IP arrangement via an OT
- Team with commercial partner via an OT
- Contracting that can avoid protests to the GAO (e.g., OTs)
- Use technical integration contract modifications or ECPs
- Contract schedule incentives
- Proven contractors only

Oversight

- Senior board of directors (direct approval or oversight to PM)
- Delegated board of directors (direct approval or oversight to PM)
- Reinforced acquisition chain of command
- Limiting the number of programs reporting to each PEO
- MDA delegated to Component Acquisition Executive (the default)
- MDA delegated to PEO
- MDA delegated to PM

NOTES: Combined approaches employ more than one approach in a cohesive model. COTS = commercial off-the-shelf; ECP = engineering change proposal; FAR = Federal Acquisition Regulation; GAO = U.S. Government Accountability Office; GOTS = government off-the-shelf; GSA = U.S. General Services Administration; IDIQ = indefinite delivery, indefinite quantity; JCIDS = Joint Capability Integration and Development System; (J)EON = (Joint) Emergent Operational Need; JROC = Joint Requirements Oversight Council; (J)UON = (Joint) Urgent Operational Need; MDA = Milestone Decision Authority; OT = other transaction; PEO = program executive officer.

There is no deterministic way to select which more-responsive acquisition approaches to use. Nevertheless, to narrow the option space, guidance that is based on the conditions of the acquisition in question can help senior, experienced personnel to determine what traditional acquisition controls are relevant and what might be unnecessary or tradable given other priorities.

Of the four selection criteria identified, the necessary conditions are, perhaps, the best starting place but also the most diverse. The variety of potentially necessary conditions (Box S.2) and the diversity of approaches that can be selected complicate efforts to sort through which approaches might be appropriate for a given situation, so we developed a simple lookup software tool that extracts potentially appropriate approaches using provided conditions (Figure S.1). The pairings are based on literature reviews, discussions with experts, and case studies. This tool cross-references program conditions with potentially suitable approaches but is dependent on the experience of the user to select which ones (or combinations) are applicable. Note that some approaches appear to be generally applicable (no conditions appear to be explicitly necessary for their use) while others have multiple conditions.

Box S.2. Range of Key Conditions to Consider in Selecting More-Responsive Acquisition Approaches

Contracting
- Contractor prequalification possible or needed
- Needs bidding approach beyond FAR
- Other Transaction Authority available
- Sole-source applies[a]
- Intellectual property protections needed beyond the FAR

Financial
- Able to influence budgets
- Budgetary resources obtainable
- Budget needs relatively small in the near term
- Funding available in prototyping account

Process
- FAR exemptions consistently applicable
- Program small dollar value

Product
- Able to use existing capabilities
- System modification, organic support, or alternative production contemplated

Provider
- Government development
- Needs strategic partnership beyond FAR

Workforce
- Contractor team highly skilled
- Contractor PM exceptionally experienced and skilled
- Empowered decisionmakers
- Formal feedback loop for training or education
- Government staff capacity
- Government staff highly skilled
- Government PM exceptionally experienced and skilled

Requirements-related
- Commonly needed contractor good or service
- Contracting areas known for future
- Explore concept of operations
- Incremental capabilities useful
- Involve users in exploring capability options
- Joint issues largely absent
- Operators can tolerate unknown risks of system failure
- Requirements: flexible level of acceptance
- Requirements might be tradable given cost, schedule, or other issues that emerge during acquisition
- Suboptimal, modular architecture and standards are sufficient
- Threats are changing
- Well-defined good or service
- Requirement priority high
- Requirements urgent or emerging (e.g., [J]UONs, [J]EONs
- Value determination needed
- Requirements approval authority at the component level

Risks (general)
- Security risks from foreign content are low
- Risks low (generally)

Technology-related
- Commercial capability or technology of interest
- Learning curve high or Infrastructure costs high[b]
- Production can be quick
- Prototyping possible in short time
- Prototype successful
- Risks of technology high [a]
- Software dominant
- Technical development needed and significant [a]
- Technology relatively mature

[a] FAR 6.302, 2019.
[b] These conditions would normally slow acquisition (or increase costs), but that fact might justify the use of an alternative approach or process that might then accelerate acquisition.

Once the necessary conditions are reviewed and a set of candidate approaches is available, the implementation considerations and stage applicability should be reviewed by decisionmakers to see which are feasible. Finally, a review of the domains covered can help yield a set of approaches designed to mitigate not only barriers in the acquisition enterprise but also effects from the requirements, financial, technical, and intelligence communities that can slow acquisition.

Figure S.1. Lookup Tool for Identifying Approaches Based on Conditions

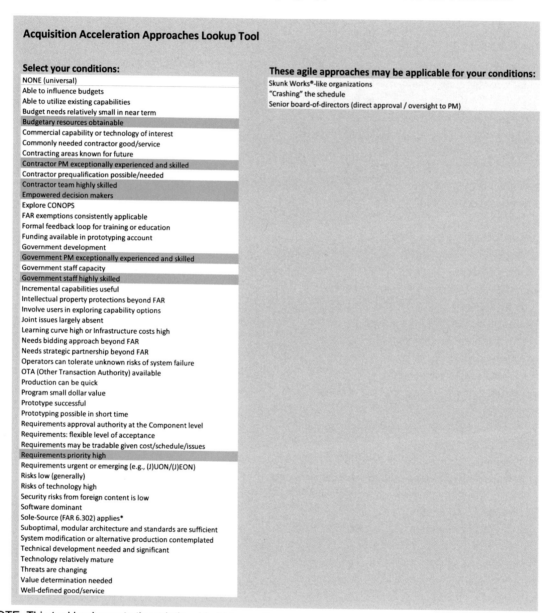

Acquisition Acceleration Approaches Lookup Tool

Select your conditions:

NONE (universal)
Able to influence budgets
Able to utilize existing capabilities
Budget needs relatively small in near term
Budgetary resources obtainable
Commercial capability or technology of interest
Commonly needed contractor good/service
Contracting areas known for future
Contractor PM exceptionally experienced and skilled
Contractor prequalification possible/needed
Contractor team highly skilled
Empowered decision makers
Explore CONOPS
FAR exemptions consistently applicable
Formal feedback loop for training or education
Funding available in prototyping account
Government development
Government PM exceptionally experienced and skilled
Government staff capacity
Government staff highly skilled
Incremental capabilities useful
Intellectual property protections beyond FAR
Involve users in exploring capability options
Joint issues largely absent
Learning curve high or Infrastructure costs high
Needs bidding approach beyond FAR
Needs strategic partnership beyond FAR
Operators can tolerate unknown risks of system failure
OTA (Other Transaction Authority) available
Production can be quick
Program small dollar value
Prototype successful
Prototyping possible in short time
Requirements approval authority at the Component level
Requirements: flexible level of acceptance
Requirements may be tradable given cost/schedule/issues
Requirements priority high
Requirements urgent or emerging (e.g., (J)UON/(J)EON)
Risks low (generally)
Risks of technology high
Security risks from foreign content is low
Software dominant
Sole-Source (FAR 6.302) applies*
Suboptimal, modular architecture and standards are sufficient
System modification or alternative production contemplated
Technical development needed and significant
Technology relatively mature
Threats are changing
Value determination needed
Well-defined good/service

These agile approaches may be applicable for your conditions:

Skunk Works®-like organizations
"Crashing" the schedule
Senior board-of-directors (direct approval / oversight to PM)

NOTE: This tool implements the relationship between the necessary conditions detailed and the acceleration approaches. The user selects or deselects conditions on the left side, and the spreadsheet tool looks up which approaches might be appropriate and lists them on the right side.

Perspectives on How to Accelerate

A common method of speeding acquisition is to prioritize schedule over cost or technical performance (i.e., to recognize the so-called iron triangle). In our analysis of more-rapid and more-agile acquisition approaches and organizations, we found that there are many more approaches to consider. In categorizing the necessary conditions for each acceleration approach, we found added perspective on accelerating acquisition.

Requirements. Being flexible and willing to make trade-offs when setting or adjusting requirements is a common way to enable acceleration. Working with users to identify acceptable levels of capabilities rather than holding up system development until all requirements are met can also speed acquisition. Other approaches can also be leveraged (e.g., streamlining the requirements processes, understanding the priorities, and iteratively exploring the schedule [and cost] implications of individual requirements as insights are gained during development and production to achieve a more informed value decision that is based on benefits versus cost and schedule).

Technology. The flip side of requirements are the technical capabilities to meet those requirements. Developing and maturing new capabilities can be very time consuming; therefore, many approaches rely on the use of mature or commercially available technology (at least for initial builds of the systems). Also, certain approaches (such as relying on experienced providers) can be helpful when technology is known to be difficult.

Budget and Cost. Other financial considerations can speed acquisition beyond simply paying more in costs to get capabilities faster. Acquiring a system requires a validated requirement, technology, and money. Without access to funds, a new requirement simply cannot be pursued. Thus, access to funds are critical for many approaches. The DoD has only partial control of budgets because Congress ultimately controls the purse strings. At the same time, the Office of the Secretary of Defense, senior military department leadership, and the financial management community within the DoD also play significant roles. This challenge is widely known, but authorizers and appropriators are loath to make flexible funds available.

Contracting and Other Processes. Simply having faster, more-streamlined processes is a commonly advocated solution for accelerating acquisition. Few processes are identified as conditions for more-responsive approaches, but there are specific approaches themselves that are process related (e.g., the seven oversight approaches, the 11 contracting approaches, special processes for [J]UONs and [J]EONs, and rapid fielding authorities).

Workforce. Perhaps the most hidden (and sometimes forgotten) enabler of rapid acquisition is a highly skilled workforce. Workforce proficiency, for example, helps to mitigate risk and enable faster acquisition. Workforce proficiency is cited as critical to such approaches as Lockheed-Martin's Skunk Works concept and the Air Force Rapid Capabilities Office. Proficiency can also lead to innovation in addressing technology maturity issues (both to solve them and to do so faster). Proficiency can allow delegation of some decisionmaking to lower levels, thus streamlining management and oversight. The importance of workforce capacity and capability is widely recognized (e.g., by the strong support in Congress and the DoD for rebuilding the government acquisition workforce).

Risk. Finally, going faster can introduce risks all around, such as system or acquisition failure, cost and schedule growth, and unintended or unknown side effects (e.g., increased higher sustainment costs; less coordinated systems). If trade-offs are not made and decisionmakers simply try to go faster, then risks are higher. Some approaches directly involve taking risks to go faster (e.g., in testing to user acceptance rather than to a full set of developmental and operational tests), but risks are often implicit or less understood in approaches and thus less discussed. Therefore, careful consideration should be taken of the possible and even unknown implications of going faster. Success in using these approaches in the past, however, shows that it can be done.

Gaps and Challenges

As evident by the diverse set of more-responsive approaches in use, the acquisition community does have ways to go faster. However, this community cannot go it alone. Fundamentally, acquisition requires three things to proceed: a validated requirement, money, and technology or services to apply. Requirements and financial resources are managed by other communities in the DoD, so those also need to proceed quickly or be tailored to readily available technology or services if acquisition is to go faster. Obtaining financial resources on short notice is perhaps the most difficult step because the budget process is conducted annually with a long lead time and Congress ultimately controls not only funding levels but also what those funds can be applied to. The DoD has long recognized the need for short-term, flexible financial resources to fill the gap, but Congress has been reluctant to allocate moderate flexible resources, even with contemporaneous ex post facto explanations of what the resources were applied to. Budgetary reprogramming can reallocate funds for critical needs, but thresholds are relatively modest and there is an overall annual limit.

In addition to there being few budgetary solutions, a simple examination of approaches by basic domains revealed very few approaches that improve the interface between acquisition and the Intelligence Community to better understand and track threat evolution (when systems must stay current against threats).[1] Ideally, the requirements community would serve this function, but requirements are often static. Acceleration approaches that involve close working relations with users and the requirements community to evolve requirements as designs and technology are explored during acquisition should help to bridge this gap, but further analysis of what the acquisition community needs from the Intelligence Community and how it can best be obtained might be warranted.

_____

[1] Note that updating requirements given evolving threats can slow the program relative to the original requirements and schedule and thus be potentially criticized for requirements creep. However, if the program is expected to stay current against evolving threats, then schedule growth will be reduced if the program obtains intelligence on evolving threats sooner rather than later.

Conclusions

Acquisition can be rapid. Speed often entails some level of compromise in cost or technical performance objectives, but it also involves other factors, such as investments in workforce expertise and experience, ready availability of financial resources, and a willingness to accept operational capabilities incrementally. Process improvements can help, but, without addressing these other factors, the gains are likely to be modest because the fundamentals of what is needed to acquire a system or service goes beyond mere bureaucratic functions.

Program conditions (size, technological complexity, schedule, etc.) are associated with what acquisition acceleration methods and strategies are applicable. It is incumbent on leaders and PMs to select the right ones for their specific programs to be successful. The DoD has used various acquisition approaches successfully for decades to improve responsiveness to warfighter threats. As technological opportunities and the pace of threats increase, the availability of a wide variety of approaches described in this report holds promise that the DoD can keep up with or even stay ahead of these threats while leveraging opportunities to retain defensive advantage during operations and in the variety of supporting functions and capabilities.

Acknowledgments

Many thanks go to Anthony Reardon, Senior Executive Service; Col Jeffrey W. Bogar; Lt Col David Tatum; Maj Patrick T. Leary; and Maj Eric J. Charest for their support, encouragement, and guidance in this research. Irv Blickstein and Brian Persons of the RAND Corporation provided very useful comments and suggestions on earlier materials for this report. The peer reviews by Lauren A. Mayer, William Shelton, and Jeffrey A. Drezner provided valuable fresh looks and helpful expert comments and suggestions on the final draft. Thanks also go to Jim Powers at RAND for his support and encouragement, including help on the user interface for the spreadsheet tool. Arwen Bicknell, our editor, helped us communicate properly and clearly. The authors claim any errors.

Abbreviations

ACAT	Acquisition Category
AFB	Air Force Base
AFI	Air Force Instruction
AFSPC	Air Force Space Command
ASuW	anti-surface warfare
C5	Consortium for Command, Control and Communications in Cyberspace
CAE	Component Acquisition Executive
CJCSI	Chairman of the Joint Chiefs of Staff Instruction
COCOM	Combatant Commander
CONOPS	concept of operations
COTS	commercial off-the-shelf
CSB	configuration steering board
DAF	Department of the Air Force
DARPA	Defense Advanced Research Projects Agency
DASD(SE)	Deputy Assistant Secretary of Defense for Systems Engineering
DAU	Defense Acquisition University
DFARS	Defense Federal Acquisition Regulation Supplement
DIU	Defense Innovation Unit
DIUx	Defense Innovation Unit Experimental
DoD	U.S. Department of Defense
DoDD	Department of Defense Directive
DoDI	Department of Defense Instruction
DPAP	Defense Procurement and Acquisition Policy
DSMC	Defense Systems Management College
ECP	engineering change proposal
EON	emerging operational need
FAR	Federal Acquisition Regulation
FBC	Faster, Better, Cheaper
FY	fiscal year
GAO	U.S. Government Accountability Office

GEO	geosynchronous
GOTS	government off-the-shelf
GPS	Global Positioning System
GSA	U.S. General Services Administration
IDIQ	indefinite delivery, indefinite quantity
IOC	initial operational capability
IP	intellectual property
IPT	integrated product team
ISR	Intelligence, Surveillance, and Reconnaissance
IT	information technology
IWN	immediate warfighter need
J&A	justification and approval
JCIDS	Joint Capability Integration and Development System
JCTD	Joint Concept Technology Demonstrators
(J)EON	(Joint) Emergent Operational Need
JRAC	Joint Rapid Acquisition Cell
JROC	Joint Requirements Oversight Council
(J)UON	(Joint) Urgent Operational Need
LMS	Lockheed Martin Space
LTPSI	long-term prime systems integrator
MDA	Milestone Decision Authority
MDAP	Major Defense Acquisition Programs
MOSA	modular open-system architectures
MR	management reserve
MTA	Middle Tier of Acquisition
NASA	National Aeronautics and Space Administration
NDAA	National Defense Authorization Act
NDIA	National Defense Industrial Association
NextGen	Next Generation
NRO	National Reconnaissance Office
OPIR	Overhead Persistent Infrared
OSD	Office of the Secretary of Defense
OT	other transaction
OTA	Other Transaction Authority

PAF	Project Air Force
PEO	program executive officer
PM	program manager
POM	Program Objective Memorandum
Pub. L.	Public Law
R&D	research and development
RCO	Rapid Capabilities Office
RDIF	Rapid Development Integration Facility
RFP	request for proposal
S&T	science and technology
SAE	Service Acquisition Executive
SAF/AQ	Assistant Secretary of the Air Force for Acquisition, Technology, and Logistics
SCO	Strategic Capabilities Office
Section 804	Public Law 114-92, Section 804
Section 806	Public Law 114-328, Section 806
SES	Senior Executive Service
SMC	Space and Missile Systems Center
SOF	special operations forces
SOF AT&L	SOF Acquisition, Technology, and Logistics
SOF AT&L-K	SOF AT&L Contracting
SV	space vehicle
TIA	Technology Investment Agreement
UON	urgent operational need
USAF	U.S. Air Force
USC	U.S. Code
USD(A&S)	Under Secretary of Defense for Acquisition and Sustainment
USD(R&E)	Under Secretary of Defense for Research and Engineering
USSOCOM	U.S. Special Operations Command
VOLT	Validated Online Lifecycle Threat

1. Introduction

Accelerating acquisition has been a priority for Congress and the U.S. Department of Defense (DoD) since at least the fiscal year (FY) 2016 National Defense Authorization Act (NDAA) (Pub. L. 114-92, 2015).[2] The DoD has used various approaches to acquire systems faster than normal (for example, see Moeller, 1979; Clark, 1993; Schoonover, 1994; McNutt, 1998; Lorell, Lowell, and Younossi, 2006; Williams et al., 2014; McKernan, Drezner, and Sollinger, 2015; and Van Atta, Kneece, and Lippitz, 2016).

This report provides an overview of those approaches and describes how their selection for specific situations should be based on conditions of use; the factors of acquisition addressed; and practical considerations for strategies, processes, and organizations. Our sponsor's focus was on space acquisition, but these approaches are applicable to all defense acquisition.

Study Methodology

Our research employed multiple methods to identify approaches for improving the responsiveness of acquisition to warfighter needs, characterize the conditions under which these approaches might be appropriate, illustrate how they might be combined, and discuss some practical considerations when implementing these approaches. These methods are as follows:

- case studies of responsive organizations
    - identification and assessment of individual approaches used
    - examination of interdependencies and practical implementation considerations
- reviews of the literature and our own personal experience
- analysis of approaches and challenges of particular interest to our sponsor
- construction of a selection framework
    - mapping application conditions to more-responsive acquisition approaches
    - identifying the involved domains
        - requirements, budgetary, technical, acquisition, and intelligence
    - identifying implementation considerations for six areas:

---

[2] For example, the FY 2016 NDAA (Pub. L. 114-92, 2015) mandated that the Secretary of Defense (i) establish a "Middle Tier of Acquisition for Rapid Prototyping and Rapid Fielding" (§804); (ii) establish an "Advisory Panel on Streamlining and Codifying Acquisition Regulations" (§809); (iii) and "establish procedures for alternative acquisition pathways to acquire capital assets and services that meet critical national security needs" (§805). The NDAA also expanded rapid acquisition authority (§803), gave the Secretary the authority to waive acquisition laws to acquire vital national security interests (§806), mandated a "report on efforts to link and streamline the requirements, acquisition, and budget processes" within the armed forces (§808), and mandated a "review of [the] time-based requirements process and budgeting and acquisition systems" (§810).

- acquisition strategies, acquisition processes, program structures, headquarters structures, requirements, and budgeting, as well as downsides and trade-offs.

The case studies helped us identify specific approaches that were created explicitly to be more agile and responsive and are used by DoD organizations. We included newer organizations to help identify new approaches that are less well known. In addition to examining approaches explicitly identified in the organizational case studies, we explored how other approaches emerge from the themes found in assessing implementation considerations. The list of approaches identified was augmented by approaches found in the literature, proposed by our sponsor, and drawn from our own personal experiences.

We then developed a selection framework that uses application conditions, domains covered, and practical considerations to guide users in selecting approaches applicable to their specific situations.

Responsive Approaches Examined

Before we begin, it is important to clarify the types of the approaches in our methodology.

Differences Between Acceleration and Schedule Control

Throughout this report, we intermingle approaches that have the potential for accelerating acquisition (i.e., reduce delivery time in absolute terms relative to the norm) with those that could help control or minimize schedule growth (e.g., minimize the amount of additional time needed to deal with a threat change). Approaches in the latter group might be directly related to program schedule, agility, or flexibility, or they might be enablers that can have an indirect effect on schedule. Some approaches can (even must) be applied early in a program rather than at some later point. Others involve activities that increase schedule (e.g., adding or changing requirements to deal with evolving threats) but seek to minimize the resulting schedule growth effects when such overall activities must be conducted. Our intent is to cast a wide net and include approaches that pursue shorter schedules (and thus faster acquisition) than would otherwise be realized in the end. For simplicity, we refer to all these variations as *more-rapid or more-agile approaches*.

New Versus Existing Approaches

To provide insights that might withstand (at least in part) the seemingly constant perturbations in acquisition policy and processes, we have both new and existing approaches in our taxonomy. Here we seek to identify approaches that have an effect on speeding acquisition rather than omitting those that are in vogue today but might be dismissed tomorrow (either formally or waived informally).

More Is Involved than the Overall Acquisition Process

Note that we do not focus simply on the typical acquisition pathways; e.g., urgent, middle tier, and tailorable traditional (Defense Acquisition University [DAU], undated-a). Instead, we

2

seek a broader set of approaches that work with, within, and around these processes to facilitate not only the acquisition system but the other systems (e.g., requirements, budgetary) that affect acquisition agility.

Why Speed Is Important

Military and political leadership continually cite space acquisition timelines as major concerns, so it is useful to briefly review the factors driving those discussions (Klimas, 2018). One factor is the commercialization of the space industry, which catalyzes the evolution of space technologies and can rapidly antiquate U.S. systems that have long cycle times from development to disposal (Chen, 2016). If the DoD intends to deploy technologies in space that ensure U.S. competitiveness, the DoD needs to reduce cycle times to keep technologies in space up to date but not go so fast as to field flawed systems.

A more alarming factor is the emerging global focus on space dominance. Foreign competitors, such as China and Russia, are investing heavily in their own space architectures, resulting in the deployment of unexpected threats and capabilities (Defense Intelligence Agency, 2017; Office of the Secretary of Defense [OSD], 2018a). As adversaries develop their space capabilities, they will have a consequential interest in protecting those assets, potentially resulting in investments for anti-satellite technologies that can hold U.S. space systems at risk. These developments—paired with the globalization of space, which allows competitors better access to foreign technologies and subsequently reduces research and development (R&D) cycle times—cause further concern. (For a discussion on the role of foreign technologies in China's space innovation capacity, see Cheung, 2016.) To respond to unanticipated threats catalyzed by globalization, and to seize the technological initiative to keep competitors unprepared, the United States has been motivated to reduce its space acquisition timelines.

Air Force Space Acquisition: Slower than Commercial Space Acquisition?

In response to perceptions that the U.S. government takes a decade or more to develop and launch its satellites, the Aerospace Corporation researched and analyzed historical satellite acquisition timelines to determine how government space acquisition cycle times compare with those in the commercial industry. Davis and Filip (2015) concluded that although there were instances of government satellites taking ten or more years to develop and launch, the average number of years to develop and launch first-of-a-kind satellites is seven and a half, and the average number of years to assemble and launch repeat satellites is just over three. Davis and Filip then compared this timeline with commercial timelines and concluded that the two are similar when comparisons are made appropriately. What allows commercial timelines to appear shorter is the fact that commercial technology development is conducted prior to beginning a commercial satellite program, and that time is therefore not included in the program's timeline. In contrast, in defense acquisition systems, government technology development time is counted as part of the program, primarily as part of the Technology Maturation and Risk Reduction phase before Milestone B but could include development performed after the earlier Materiel Development Decision. When timelines are defined from the start of development to launch for first-of-a-kind satellites (or start of assembly to launch for repeat satellites), durations of commercial and government satellite acquisitions are similar (Davis and Filip, 2015).

Note one important caveat, however. An alternative definition of the *acquisition timeline* is the duration between the validation of the capability requirement and initial operational capability (IOC) shortly after launch, and this is the definition that should arguably matter most in Department of the Air Force (DAF) pursuit of a responsive and resilient space architecture. With this new definition, the commercial acquisition timeline is shorter by years, but only because the commercial sector performs technology development before making a product available to customers to request. The head start on technology development allows the commercial sector to be better prepared to address customer demand when it arrives. The DAF can more closely replicate this ability through a healthy science and technology (S&T) budget. In Chapter 4, we offer insight on how the DAF can be prepared for the challenge of defending its S&T budget items.

What Can Slow Acquisition Down

Before we dive into approaches and trade-offs that can make acquisition more responsive, we describe some factors that tend to slow the system down. These might seem obvious considering what the approaches address, but it is worth discussing some key factors more explicitly. An entire report can easily be written discussing these factors, but here are a few to help illustrate the challenges.

First, recall that acquiring a capability involves more than just the acquisition enterprise. Government acquisition officials must have both a validated requirement and financial resources before acquiring something. Also, requirements are often driven by threats, so intelligence is needed. Finally, requirements are often addressed through technical means, so mature technologies that can be relied on are also needed. The process for obtaining all these (requirements, money, intelligence, and technology) must be expedient and responsive, but the systems that manage these domains all have their own bureaucracies and processes that can be slow or unresponsive. For example, programs are often blamed for not being responsive to changing threats (i.e., changing requirements to address new situations). Conversely, when programs do change requirements, they are blamed for creep and associated cost and schedule implications. There are natural tensions between serving narrower military service needs and broader joint needs; this tension leads to desires to bypass joint processes or oversight authorities, on the one hand, and processes that can slow acquisition, on the other.

The acquisition enterprise itself has its own processes, but they are not the only potential source of slowness. Workforce incentives often punish realized risks, so conservatism prevails. Staff in the bureaucracy also tend to want to help (or feel responsible for helping) programs, so micromanagement in oversight is a concern. Workforce expertise is mixed, so oversight is still needed but consumes time and resources.

There is also a tension between rapid fielding using mature technologies and pushing the state of the art with new, immature technologies. Rapidly produced systems also tend to have compromises not only in performance relative to need but also in reliability and sustainability.

The Basics: Cost, Schedule, and Performance

It is useful to remember that going faster usually has consequences in terms of trade-offs and risk. The trio of cost, schedule, and performance is often called the *iron triangle* because history has shown that dramatic improvements generally cannot be made in all three areas at once (i.e., decisionmakers must choose what to prioritize). Attempting to do so often incurs higher risks in one or all of these areas. Thus, accelerating acquisition is all about deciding what is more important. It is possible to move faster, but doing so usually entails higher cost or compromises on technical performance.

For example, actively seeking satisfactory (but not exceptional) capabilities during development involves actively exploring the trade space that provides rapid, cost-effective solutions. These so-called *80-percent solutions* involve accepting lower performance (often both initially and during development) as the developers push the state of the art and find the sweet spots (the pivot point of diminishing returns for added work). Engineers often do not know in advance where these spots are, so this approach benefits from user feedback and an ability to get requirements relief during development.

Some concepts seemingly try to break the iron triangle but, on closer examination, merely illustrate its operation. One example is the National Aeronautics and Space Administration (NASA)'s Faster, Better, Cheaper (FBC) concept (Spear, 2000), pushed in the 1990s by NASA's then-Administrator Daniel Goldin. FBC was abandoned in the early 2000s after two Mars mission failures (the Climate Observer and the Polar Lander) and four other mission failures out of 25 missions (Oberg, 2000; Inspector General of the National Aeronautics and Space Administration, 2001). A close examination of the FBC concept reveals that it is misleading to say that NASA was trying to break the iron triangle. NASA did want to go "Faster" and "Cheaper," but "Better" did not mean there was no compromise in performance. Essentially, NASA traded grander missions and performance objectives for smaller systems that were more cost-effective with significantly smaller budgets. Whether NASA obtained "better" results depends on perspective and metrics. For example, the concept was essentially canceled after the high-visibility failures mentioned above. However, Ward (2010) counters that, out of 16 FBC projects, NASA had ten successful missions, all of which combined cost less than the Cassini mission to Saturn (i.e., ten FBC successes cost less than one traditional mission). Also, Dillon and Madsen's analyses (2014, 2015) suggested that, by some measures, the 16 FBC missions were more efficient at providing insights than was a large-scale mission, such as Cassini. The

FBC example, therefore, might accurately illustrate the concept of trade-offs and the importance of clearly articulating objectives and priorities.

What Is Different About Space Acquisition

Space acquisition has some relevant differences from other types of military system acquisition. These differences provide relevant context when considering approaches for space acquisition programs. Here, we separate these differences according to whether they are external or internal to space acquisition.

First, space acquisition occurs in a dynamic environment characterized by external emerging threats, a commercializing industry, and an unpredictable organizational environment. These external influences can shape the DAF's decisions on how to craft acquisition strategies and choose which approaches to use. Emerging threats, for example, are unique to space in that most other domains (aside from cyber) are not fighting an active, nonkinetic war. The continuous nature of these emerging threats should be taken into consideration for decisions on approaches. For example, the consistent presence of new threats and the fact that solutions are likely to become obsolete more quickly in the space domain might reinforce the need to acquire 80-percent solutions rather than the perfect 100-percent solution. The space sector (like the cyber sector) continues to see an influx of commercial companies and nontraditional defense contractors. Traditional defense space contractors are responding to this influx through corporate adjustments and realignments to be competitive.

Second, in terms of approaches, this new and evolving space economy could provide the DAF with innovations and capabilities to exploit but also might introduce more risks to manage. We discuss further details on this in our section on industry perspectives of DAF space acquisition.

Finally, the organizational environment for space operations and acquisition is unpredictable given ongoing proposals for changes (e.g., creating a new Space Force). Choosing an approach might be agnostic to this factor, but, depending on the situation and program, there might be a need for acquisition strategies to use approaches that can deliver more consistency through a turbulent organizational environment.

In addition to these external factors, space acquisition involves several internal factors that make it more complicated. First, space systems have many stakeholders that do not understand the technical details of space systems but rely on their capabilities. Therefore, the Joint Capability Integration and Development System (JCIDS) process can be troublesome. The DAF might be inclined to seek to bypass the JCIDS process (which can add delays with its additional reviews). However, JCIDS assesses whether joint military capabilities are needed to efficiently meet applicable requirements in the National Defense Strategy (OSD, 2018b) by identifying, approving, and prioritizing gaps in such capabilities. Thus, JCIDS is a forum for informing cost, schedule, and system performance trade-offs from a joint perspective; deciding whether to pursue similar capabilities as a joint system; and resolving stakeholder differences on requirements for joint capabilities.

Second, satellites in orbit are largely unable to be modified or upgraded, although broader capabilities for on-orbit servicing, reprogramming, and reconfiguration are in development. Because upgrades are limited, users and acquirers are inclined to push for the 100-percent solution. Similarly, on the launch side, acquirers have one shot for a successful launch. Therefore, launch acquirers are inclined to be risk averse, imposing mission assurance requirements and government oversight on launch service provider operations. These factors can create organizational inertia that influences the debate on approaches.

Third, space acquisition requires orchestrating multiple interdependent space programs: To effectively deliver warfighting capability in space, the DAF must synchronize and harmonize satellite, launch, and ground acquisition programs. Depending on the mission, for example, launch service and satellite acquisitions must be intimately connected for a period of time before launch. Launch service providers are often on fixed-price contracts to execute all the technical analyses (e.g., coupled loads analysis), preparation and execution of satellite and launch vehicle integration, launch rehearsals, etc. Any program perturbations, such as requirements or funding changes, can spell disaster for meeting the satellite launch date and successfully launching the satellite. Thus, the DAF will need to consider these complicated relationships when determining the appropriate acceleration approach.

Current Events Affecting Space Acquisition

The concerns about space acquisition timelines have resulted in several efforts to expedite acquisition—notably, the recent plan to establish a Space Force by 2020. The Space Force initiative is intended to reduce bureaucratic steps traditionally associated with DAF acquisition to allow prioritized, efficient, and rapid space acquisition that is responsive to emerging foreign threats. The proposed Space Force will also feature a Space Development Agency responsible for both the development and acquisition of space systems (Insinna, 2018). The Space Force proposal is a contentious one, with opponents arguing that it insufficiently addresses the hurdles within space acquisition and that the DAF is primed to handle U.S. space acquisition needs (Bender and Klimas, 2018).

A second significant effort within the DAF is a revamp of its space acquisition process that features a reorganization effort of the Space and Missile Systems Center (SMC) into "SMC 2.0." This center is a subordinate unit of Air Force Space Command (AFSPC) and bears much of the space acquisition burden for the DoD. The SMC 2.0 effort is a broad reorganization intended to accelerate acquisition through the inclusion of a chief architect intended to provide enterprise-level guidance and three program executive officers (PEOs) to reduce stovepiping across the organization (Erwin, 2018a; Underwood, 2018). Although these are significant changes, it might take years to reap broad benefits across space acquisition.

Another major event affecting space acquisition is the establishment of the Space Rapid Capabilities Office (RCO) at Kirtland Air Force Base (AFB) (Pub. L. 115-91, §1601, 2017; 10 USC §2273c). The Space RCO replaces the Operationally Responsive Space Office and is modeled after the DAF RCO.[3] It is intended to handle the most-urgent space acquisition needs. Two other events with implications for space acquisition were also implemented by the FY 2018 NDAA. The first of these terminated the A-11 position of Deputy Chief of Staff of the Air Force for Space Operations (Pub. L. 115-91, §1601(b)(1)(D), 2017), which was responsible for transitioning and normalizing space operations as a warfighting function and for providing combatant commanders with required capabilities. The other change shifted Milestone Decision Authority (MDA) to the Commander of AFSPC for defense space acquisition (Pub. L. 115-91, §1601(a), 2017; 10 USC §2279c(b)(2)), intending to streamline the acquisition process. Although it is unclear whether these changes will reduce acquisition timelines, the efforts signal a strong priority from both DoD and civilian leadership that space acquisition cycle times need improvement.

Note that it is beyond the scope of this study for us to assess whether the creation of the Space Force will reduce cycle times. Furthermore, the details of the new Space Force organization are still in flux, making it hard to make such assessments.

Industry Perspectives on Space Acquisition Agility

Industry's perspectives can help us better understand industry capabilities to support agile DAF space acquisitions. Here, we discuss some industry perspectives from both old and new companies. The *old* companies are traditional defense contractors, such as Lockheed Martin Space (LMS) and the United Launch Alliance (a joint venture between LMS and Boeing Defense, Space & Security). The *new* companies are the nontraditional defense contractors, such as SpaceX and Blue Origin. Note that we use the terms *traditional* and *nontraditional* somewhat loosely compared with the specific definitions found in the Defense Federal Acquisition Regulation Supplement (DFARS) (2019) and 10 USC §2302(9), which define *nontraditional* essentially as a company with no DoD contract or subcontract. We assess that a traditional defense contractor has an established history of doing space business with the DAF and is also likely reliant on this business to close its business case. On the other hand, a nontraditional defense contractor is generally new to DAF space business or might have a small share of DAF contracts and is likely not reliant on this business to close its business case. We selected two companies, LMS and Blue Origin, to capture some perspectives of both a traditional and a nontraditional defense contractor (Box 1.1).

Disruptive competitors and competition can also drive innovation and agility. The LMS perspective exemplifies how traditional defense contractors are beginning to offer more-agile solutions (Ambrose, 2018a; Ambrose, 2018b; Ambrose, 2018c; Ambrose, 2018d; Erwin, 2018c).

---

[3] DAF RCO and Space RCO are described in greater detail in Chapter 4.

The DAF is not solely driving this, however. These contractors are adapting to both a change in DAF space strategy and the commercialization of space. First, DAF space strategy fostering a more robust, responsive, and resilient space architecture drives these traditional contractors to offer more diversity and affordability in their products and services. Second, the commercialization of space has resulted in more nontraditional competitors in the government sector and more opportunities to gain market share in commercial space sectors. Competition in both sectors naturally drives more diversity and affordability in products and services. Together, these two effects propel traditional defense contractors to reassess their business strategy, including their manufacturing processes, products or services offered, and price points. From the LMS perspective, we see that these effects are doing just that. On the launch side, United Launch Alliance (a traditional defense contractor) is undergoing a similar transformation with the pursuit of its Vulcan rocket. We expect that these transformations could support more-responsive acquisition opportunities for the DAF.

Box 1.1. Selected Commercial Industry Perspectives Toward Rapid Space Acquisition

LMS[a]

- Revamping manufacturing capabilities to accommodate all satellite sizes and higher production throughput
- Looking to halve satellite cost and delivery time
- Pursuing commonality in manufacturing processes and technology or hardware among satellite variants
- Forward looking in satellite technologies, such as artificial intelligence and reprogrammability, that can benefit all customers
  - "Technologies in the new space economy will . . . have to advance the goals of both commercial and government customers. Best practices should be shared and new ideas need to spread quickly."
- Speed, in addition to cost and quality, advantages can be achieved through synergies between commercial and government products

Blue Origin[b]

- Pursuing a single big success that is worth thousands of failed ventures:
  - "Winners pay for thousands of losers."
- Proactively seeking solutions to customer needs, even before they ask or think they want it (e.g., no customer asked for Amazon's Echo speaker with Alexa, yet this invention is now in high demand)
- Using commercial solutions whenever possible:
  - "Reserve your custom requirements for things where you really need special sauce."

[a] Ambrose, 2018a; Ambrose, 2018b; Ambrose, 2018c; Ambrose, 2018d; Erwin, 2018c.
[b] Boyle, 2018.

The Blue Origin perspective exemplifies how nontraditional defense contractors have been setting the pace and expectations for the space industry and how their products or services (which are relevant to DAF space strategy) can directly contribute to more-responsive acquisition (Boyle, 2018). However, nontraditional contractors argue that acquisition agility is specifically hindered by government-unique requirements. Their business hinges on their competitiveness in the market that brings in the most profit (which might not be in the government marketplace), which means that if they deviate too much from their core business practices to accommodate government-unique requirements, they could lose profit and potentially risk the future of their entire business. If, however, there are common requirements, then the results can be mutually beneficial. The DAF can also benefit because the commercial products or services are driven by the expectations of numerous, demanding commercial customers and are therefore affordable and timely. Commercial solutions are also upgraded on a regular basis, meaning some updates or refreshes do not need to go through the government acquisition process.

The DAF could better capitalize on agile opportunities through a holistic and program-by-program analysis of government-unique requirements. There might be opportunities for the DAF to intelligently reduce these requirements in situations in which it makes sense to facilitate greater use of the commercial space sector. This will inarguably expand the DAF's solution space, allowing the DoD greater latitude in achieving a more robust, responsive, and resilient space architecture.

We caution that the DAF needs to manage the risks associated both with commercial solutions and with defense contractor partnerships in this period of space commercialization. First, commercial solutions might provide short-term rather than long-term competitive advantages because adversaries have greater insight to replicate these capabilities. Thus, space superiority might be at risk if the DAF is overly dependent on commercial solutions. Additionally, commercial solutions are likely to present increased supply-chain risks, such as counterfeit parts and vulnerabilities to cyberattacks.

Second, partnerships with defense contractors in this period of space commercialization might be at risk because these contractors' incentives and motivations are more commercially focused; this is certainly the case for nontraditional defense contractors now, but it might also be true for traditional defense contractors in the future. A lack of long-term partnerships could result: Some of these companies might not need government work to close their business cases, so there might not be an incentive to continue doing business with the government when the return on investment is no longer attractive. This risk could also be realized if these defense contractors go bankrupt, which is not an unlikely outcome in the competitive space economy with high-risk endeavors. This partnership risk is one reason why the DAF is interested in the concept of a long-term prime systems integrator (LTPSI), which is discussed in detail in Chapter 4.

The DAF's unique challenge will be balancing the opportunities and risks of commercial solutions, determining which programs and capabilities need government-unique solutions, and working with industry on finding the most effective amount of custom technology for each one, all while ensuring that a healthy mix of industry partners is invested with the DAF in the long term.

Report Outline

This report facilitates the identification and selection of appropriate acquisition approaches in a logical sequence: First, we discuss approaches and their use conditions (Chapter 2), methods to facilitate selection (Chapter 3), and examples of organizations that integrate and employ multiple approaches (Chapter 4). Readers can skip to subsequent chapters if their interest is more on selection techniques or organizational models than details on specific approaches.

Chapter 2 summarizes the variety of approaches uncovered by our analysis of rapid acquisition organizations and other approaches from the literature and our experiences. The applicability of individual approaches to a given situation is illustrated by our identification of key conditions necessary for the use of each approach and the practical implementation considerations—i.e., their effects on acquisition strategies, acquisition processes, program structures, headquarters structures, requirements, and budgeting, as well as downsides and trade-offs. (Readers who want to focus on selection mechanisms might skip this long chapter and use it later as a reference.)

Chapter 3 provides mechanisms for selecting applicable approaches in practice. We describe a software tool we constructed that uses conditions to identify candidate approaches. Those candidates can then be reviewed against (1) the implementation considerations from Chapter 2, (2) a new table identifying the domains that each approach involves (i.e., requirements, budgeting, technical, acquisition, and intelligence), and (3) whether the candidates are appropriate for the stage that the acquisition in question is in.

Chapter 4 discusses some organizational models that illustrate how approaches can be combined, along with discussion of the practical implementation considerations of those organizational models and the domains covered.

Chapter 5 offers an overarching discussion and conclusions.

Appendix A provides additional insights into practical implementation considerations for three approaches.

Appendix B discusses additional practical considerations of recompeting or keeping an LTPSI.

Appendix C provides initial ideas on how the budgets for S&T feeds might be protected or stabilized as a way to mature technology and speed the subsequent acquisition.

Appendix D provides some discussion on the challenges in using simpler decision charts to help select candidate approaches.

2. Approaches for More-Responsive Acquisition

In this chapter, we review and assess a variety of new, existing, and historical approaches for accelerating acquisition. Our research revealed no good source with a comprehensive list of approaches, let alone discussion and ways to select which approach might be appropriate for a given situation. In addition to asking us to review new and existing approaches, the DAF also asked us to examine, in particular, whether retaining prime system integrators on space programs in the long run might be a new consideration.

Responsive Acquisition Approaches

Many approaches already exist for more-responsive acquisition. Box 2.1 lists the approaches we identified and analyzed, grouped into categories.[4] These approaches were identified through examination of (1) those used by selected rapid acquisition organizations, (2) recent legislation, (3) the published literature, and (4) our own experiences. The rapid acquisition organizations we examined are

- Big Safari (Harrington, 2018; 88th Air Base Wing Public Affairs, 2018)
- acquisitions by the DoD and the U.S. Air Force (USAF) Special Operations Forces (SOF) (U.S. Special Operations Command, Special Operations Forces [Acquisition, Technology, and Logistics], undated)
- the National Reconnaissance Office (NRO) (undated-a)
- the DAF RCO (USAF, 2009)
- the Joint Rapid Acquisition Cell (JRAC) in OSD (DAU, undated-b)
- the Army's Consortium for Command, Control and Communications in Cyberspace (C5)
- the Defense Innovation Unit (DIU) (undated)
- the Defense Advanced Research Projects Agency (DARPA) (undated-b).

These organizations are discussed in Chapter 4; we also illustrate how individual approaches can be integrated in practice.

Other organizations could have been included, but this set generated a reasonably comprehensive set of approaches and spanned key portions of the acquisition domain. Organizational analysis consisted of literature reviews, selected meetings with representatives of these organizations, and reviews of recent legislation related to these organizations.

---

[4] Note that we do not focus simply on the typical acquisition pathways—e.g., urgent, MTA, and tailorable traditional (DAU, undated-a)—but seek a broader set of approaches that work with, within, and around these pathways, including not only the acquisition system but also other systems (e.g., requirements and budgetary) that affect acquisition agility.

We also sought approaches that address issues from domains outside acquisition (such as addressing factors in the requirements, budgetary, and intelligence communities that can slow acquisition).

Box 2.1. Approaches for More-Responsive Acquisition, by Category

Higher-Level
- Acquisition tailoring
- Technical risk mitigation

Combined
- Skunk Works–like organizations
- "Agile" development

General
- In-house engineering and production
- Formal, active lessons learned process
- Standing FAR waivers

Intelligence-Related
- Faster intelligence

Requirements-Related
- Requirements: partial (80%) solutions acceptable to users
- Requirements: keep stable (avoid "creep")
- CSBs
- Bypass JCIDS and JROC

Schedule-Based
- Crashing the schedule
- Fast-tracking (parallelization; concurrency)
- Streamlining
- Focused work
- Work-breakdown schedule reviews
- Schedule reserve (margin; contingency)
- Readiness-based gates or milestones
- Hard, fixed schedules

Financial-Related
- DoD Rapid Prototyping Fund
- Reprogramming
- Budget stability measures
- MRs

Technology Development–Related
- Operational prototyping: new platforms
- Operational prototyping: middle tier of acquisition (MTA) rapid prototyping (10 USC §2302 Historical and Revision Notes; Pub. L. 114-92, §804, [hereafter referred to as "Section 804"], 2015)
- Operational prototyping: components
- Operational prototyping: components or technology (10 USC §2447; Pub. L. 114-328, §806, [hereafter referred to as "Section 806"], 2016)
- Joint Concept Technology Demonstrations (JCTDs)
- Prize competitions

Reuse
- Modifications to existing platforms
- Repurposing
- Off-the-shelf: COTS or GOTS
- Adaptation of foreign technology

Maturity-Based Modification
- Incremental acquisition
- Continuous component upgrade as technology matures

Box 2.1—Continued

Design-Related

- MOSAs
- Disaggregated architectures
- Own the technical data (especially interfaces)

Testing-Related

- Test to acceptance
- Operational testing in actual operations

Rapid-Fielding

- Rapid fielding: MTA (10 USC §2302 Historical and Revision Notes; Section 804, 2015)
- (J)UONs or (J)EONs

Contracting-Related

- Indefinite-delivery contracts (e.g., IDIQs)
- LTPSI
- Noncompetitive contracting: J&A
- GSA schedule
- Marketplaces
- Bids open to consortium members only
- Sole-source production following successful prototypes
- Custom IP arrangement via an OT
- Team with commercial partner via an OT
- Contracting that can avoid protests to the GAO (e.g., OTs)
- Use technical integration contract modifications or ECPs
- Contract schedule incentives
- Proven contractors only

Oversight

- Senior board of directors (direct approval or oversight to PM)
- Delegated board of directors (direct approval or oversight to PM)
- Reinforced acquisition chain of command
- Limiting the number of programs reporting to each PEO
- MDA delegated to Component Acquisition Executive (the default)
- MDA delegated to PEO
- MDA delegated to PM

NOTES: Combined approaches employ more than one approach in a cohesive model. COTS = commercial off-the-shelf; CSB = configuration steering boards; ECP = engineering change proposal; FAR = Federal Acquisition Regulation; GAO = U.S. Government Accountability Office; GOTS = government off-the-shelf; GSA = U.S. General Services Administration; IDIQ = indefinite delivery, indefinite quantity; IP = intellectual property; J&A = justification and approval; (J)EON = (Joint) Emergent Operational Need; JROC = Joint Requirements Oversight Council; (J)UON = (Joint) Urgent Operational Need; MOSA = modular open-system architectures; MR = management reserves; OT = other transaction; PM = program manager.

The goal is to facilitate selecting approaches appropriate for a program or organization, especially if these decisions can be made based on an understanding of what possible approaches might be appropriate. Thus, in this chapter we assess two factors for each approach:

- the necessary **conditions** for applying the approach
- practical **implementation considerations** (i.e., how acquisition strategies, acquisition processes, program structures, headquarters structures, requirements, and budgeting need to be adjusted or aligned to the approach and to account for downsides and trade-offs).

In Chapter 3, we provide frameworks for sifting through these numerous considerations when selecting appropriate approaches to use in a given situation. At that point, we assess two additional factors:

- the **domains** involved (especially requirements and financial)
- the **stage** in an acquisition's life cycle at which the approach could be used.

Applicability of Individual Approaches

Conditions for Use

In analyzing the approaches, we found that most have at least one key condition that inherently enables the application of the approach. These conditions also could relate to the inherent risk and functional trade-offs associated with the approach. For example, oversight authority might be delegated down to lower levels only if programs are smaller, pose lower risk, or both. If the delegation is deep—to a PM, for example—then we would want those PMs to be exceptionally experienced and skilled, the programs to be relatively small (meaning fewer dollars at risk), and risks generally to be low. Otherwise, some level of supervisory review (by at least a PEO, for example) would be prudent.

Following the short descriptions of each approach, a table identifies the necessary conditions we identified from a fundamental, theoretical analysis of each approach, along with a short justification of our reasoning. In each case, the methodology indicates that each necessary condition should be true for the approach to be applicable. In other words, it is the union of the conditions, not just the existence of any one of them (i.e., "and," not "or"). These judgments are based on both our expertise in acquisition and our insights into each approach.

Note that these necessary conditions do not mean that other conditions are not useful, advisable, or desirable. For example, maintaining highly skilled PMs is a generally good practice, so one might expect that to be a condition for using any approach. However, the framework we constructed focuses on conditions that are usually necessary for the approach to be applied. Most approaches will work (albeit not as well) with less experienced PMs, and many approaches have management and oversight mechanisms to deal with less experienced PMs. Approaches for which MDA is delegated down to the PM make a highly skilled PM a necessary condition, so those are the ones we identified as needing PMs with exceptionally high experience and skill levels.

Implementation Considerations

In addition, each approach has practical considerations for implementation. To help inform the acquisition selection and implementation, we summarize here our analysis of these key considerations for each approach:

- acquisition strategies
- acquisition processes

- program structures
- headquarters structures (e.g., the Air Staff and the Assistant Secretary of the Air Force for Acquisition, Technology, and Logistics [SAF/AQ])
- requirements
- budgeting
- downsides and trade-offs.

This discussion reflects our analysis and expert understanding of each approach and how its construct probably affects these areas in consideration of acquisition fundamentals and theory.

These practical considerations are brief and intended to help characterize and differentiate the approaches. Further in-depth planning and analysis (as well as coordination and consultation with general counsel) are needed before adopting and implementing these approaches.

Along with the discussion of these practical considerations for each approach, we also provide a brief discussion of some major downsides and trade-offs. As with the practical considerations, these are necessarily brief because many can involve whole reports and volumes of discussions and nuances, but these are provided to give high-level perspective and to illustrate that each approach often comes with downsides that limit application.

Higher-Level

Descriptions

Acquisition tailoring. Tailoring adjusts and customizes the acquisition process using the specific properties, risks, and needs of the acquisition in question. Critical thinking about the acquisition in question—along with knowledge about what approaches are appropriate given the specific combination of acquisition conditions—is required to tailor acquisition and choose a combination of approaches that enables accelerated cycle times in light of other strategic considerations and risks. For example, tailoring might eliminate noncritical steps in the acquisition process and allow milestone approvals without waiting for document delivery or explicitly undertaking more-risky approaches. Tailoring can consist of one or more of the approaches discussed below, including acquisition strategy elements for technology development, contracting, and oversight. This can be thought of as the ability to employ a "buffet style" of program management. In a sense, this report is all about tailoring the process for a particular acquisition rather than rote application of a default process that might contain inappropriate or unnecessary approaches and elements.

Technical risk mitigation. A wealth of approaches can be employed to mitigate technical risks (and thus their implications for schedule risk). Common approaches are reducing technical risks early (e.g., before Milestone B) or employing other risk mitigation approaches, such as parallel pursuit of alternative technologies or designs. The *Department of Defense Risk, Issue, and Opportunity Management Guide for Defense Acquisition Programs* (Deputy Assistant Secretary of Defense for Systems Engineering [DASD(SE)], 2017) provides extensive discussion of types of approaches, such as risk acceptance, monitoring, avoidance, transfer, control, and reduction. We provide other examples as specific approaches below: Some involve spending additional time up front to save much more time later when subsequent design or production can go faster. Also, pursuit of alternatives would save time if the primary technical approach fails.

Conditions for Use

Table 2.1 lists the key conditions generally necessary for use of each approach and a short discussion of the reasoning behind the conditions.

Table 2.1. Key Conditions: Higher-Level

Approach	Necessary Condition(s)[a]	Reasoning
Acquisition tailoring	• None (universal)	Tailoring is essentially universal, but we note that it requires PMs and their supporting staff to be fairly experienced and knowledgeable about acquisition fundamentals beyond rote application of predefined acquisition approaches. Prudent tailoring generally requires a PM who understands the system in question, its properties (e.g., maturity, difficulty in development), and how the general acquisition approach can be tailored accordingly. Risk decisions need to be made prudently with the potential to adjust if risks come to fruition.
Technical risk mitigation	• High technological risks • Technical development is necessary and significant	Common approaches are reducing technical risks early (e.g., before Milestone B) or employing other risk mitigation approaches, such as parallel pursuit of alternative technologies or designs. Technical risk reduction is particularly important when technical risks are high and significant development is needed.

[a] Generally, each approach needs all the features listed (not just one).

Implementation Considerations

Acquisition Tailoring

Acquisition strategies. All acquisition strategies should lay out how the process will be tailored and why.

Acquisition processes. Tailoring the acquisition process is a key overarching objective and directive in Department of Defense Instruction (DoDI) 5000.02. The process itself is adjusted to the acquisition at hand.

Program structures. Tailoring could affect which program elements are needed (e.g., whether staff are needed to prepare for a certain milestone review).

Headquarters structures. Tailoring might affect headquarters review and oversight processes and approvals.

Requirements. Unaffected.

Budgeting. Unaffected.

Downsides and trade-offs. Tailoring requires some acquisition expertise in structuring an appropriate process and balancing risks, given relative leadership priorities, among cost, schedule, and performance for the acquisition at hand. Leadership also needs to provide support for tailoring so that PMs and staff do not default to the most conservative approach possible. This might also imply employment of other methods, such as actively managing risk, raising problems early, and knowing when to cancel or curtail an effort that is not working out.

Technical Risk Mitigation

Acquisition strategies. Strategies should identify technical risks, their potential implications for program execution and schedule, and which types of risk mitigation approaches will be adopted (DASD(SE), 2017).

Acquisition processes. Specific considerations for acquisition processes depend on the risk mitigation approach (or approaches) employed.

Program structures. Needed changes in a program's structure (e.g., program elements and capabilities) depend on the risk mitigation approach (or approaches) employed.

Headquarters structures. Unaffected, although some mitigation approaches could involve the creation of large, strategic investments or the creation of entities to reduce risks and to mature technology.

Requirements. Effective structures and processes should be in place to enable the requirements community to understand the risks of acquiring desired capabilities.

Budgeting. High risks might lead to increased budgetary needs.

Downsides and trade-offs. Depending on the risk mitigation approach employed, costs might be higher (e.g., if multiple technologies are pursued in parallel), or the initial delivery of capabilities to the warfighter might be slow if risk tolerance is set too low.

Combined

These are approaches that integrate multiple elements into a combined approach. Note that organizational models combine multiple approaches and illustrate how they can be combined in real-world applications. We provide only one organizational model (Skunk Works) because it is so well known and has a well-defined set of necessary conditions. Further work could expand the taxonomy and look-up tool to include these organizational models as options to mirror.

Descriptions

Skunk Works–like organizations. This is a long-standing approach that leverages approaches for producing advanced technologies quickly (Lockheed Martin, undated; Miller, 1995).[5] This model consists of 14 key elements, including a senior board of directors, acquisition tailoring, stable requirements, and partial solutions acceptable to users. Chapter 4 discussed this model in greater depth. Generally, these approaches ensure oversight and attention to cost and

---

[5] Other organizations have emulated parts of Lockheed Martin's Skunk Works concept, such as Boeing's Phantom Works (Werner, 2012) and NASA's Swamp Works (Makufka, undated). Our report focuses on Skunk Works as an example.

schedule while streamlining design work and access to empowered decisionmakers. Historic examples of programs produced under the Skunk Works model are the U-2 and SR-71.

"Agile" development.[6] This approach is primarily applied to software development under which requirements and solutions evolve through the collaborative effort of self-organizing and cross-functional teams and their customers or end users. It emphasizes incremental development and continual improvement. The basic idea is to develop increments of capability in *sprints*—repeated, short time frames—that allow users to evaluate and provide feedback for the development of the next increment. The approach allows the developers and users to explore the design space, quickly deliver initial usable capabilities, and identify the most-effective solutions in terms of cost and time (for example, see Randell and Zurcher, 1968; Gilb, 1977; Zultner, 1988; Dana, 1993; Larman and Basili, 2003; Lapham, Williams, et al., 2010; Northern et al., 2010; Reagan and Rico, 2010; Reagan and Rico, 2010; Lapham, Miller, et al., 2011; GAO, 2012b; Markuson and Flasher, 2014; Modigliani and Chang, 2014; Oar et al., 2015; Miller et al., 2016; Camm et al., forthcoming; and Kim et al., forthcoming). Chapter 4 also provides examples of organizational models that integrate multiple approaches into an integrated formula for more-responsive and more-agile performance.

Conditions for Use

Table 2.2 lists the key conditions generally necessary for use of each approach and a short discussion of the reasoning behind the conditions.

Table 2.2. Key Conditions: Combined

Approach	Necessary Condition(s)[a]	Reasoning
Skunk Works–like organizations	• Obtainable budgetary resources • Highly skilled contractor team • Exceptionally experienced and skilled contractor PM • Highly skilled government staff • Exceptionally experienced and skilled government PM • High priority • Empowered decisionmakers	Although the full list of elements in the original Skunk Works concept is longer (Lockheed Martin, undated; Miller, 1995), key elements are (1) highly skilled and thus trusted teams on both the government and contract sides and (2) high-priority efforts enabling quick government decisions.
"Agile" development	• Flexible requirements • Predominantly software content • Users involved in exploring capability options	The "Agile" development approach was originally developed for software, although parts of the approach can be adapted to some hardware applications. One key element is developing increments of software on short cycles (weeks) for users to evaluate and provide feedback on needs and next steps. This inherently involves a degree of flexibility in requirements, allowing users to help determine what level of capability is good enough through a dynamic exploration of what is easier to produce quickly (GAO, 2012b).

[a] Generally, each approach needs all the features listed (not just one).

---

[6] "Agile" development is the prescribed name of a specific approach. We have used quotation marks and capitalization to distinguish this specific approach from other approaches that can enable acquisition agility.

Implementation Considerations

Skunk Works–Like Organizations

Acquisition strategies. The strategy focuses on the set of operating principles, including the 14 key points from the original Lockheed Skunk Works concept (Lockheed Martin, undated; Miller, 1995).

Acquisition processes. Processes focus on rapid review, selection, and funding of prototypes and proposed systems.

Program structures. Programs are focused on supporting the Skunk Works and Phantom Works processes, with close oversight and reviews of contractor efforts, ensuring that the government and contractor program offices are staffed with highly capable people who have access to leadership who can make quick decisions on requirements, funding, and program advancement.

Headquarters structures. Decisionmakers (MDA, funding, and requirements) need to be readily available to review and make quick decisions to advance the program.

Requirements. Key (even aggressive and revolutionary) requirements can be fixed, but often other requirements must be flexible to enable rapid achievement of the key capability of interest.

Budgeting. Funding needs to be available and somewhat flexible to enable rapid decisions and pursuit of revolutionary capabilities.

Downsides and trade-offs. Limitations as outlined in the necessary conditions are, perhaps, the biggest trade-off of this approach. Funding and design decisions need to be made by (often very senior) empowered individuals, and even their congressional oversight and approval can limit their ability to make quick decisions. Also, the approach depends on having extremely capable leaders and teams on both the government and contractor sides to speed work and empower trust; such expertise is limited.

"Agile" Development

(A more detailed discussion can be found in Appendix A.)

Acquisition strategies. "Agile" development is a major element of the strategy. Contracts must have flexible requirements that focus on services delivered during development rather than on fixed product requirements. End state is determined through user feedback on incremental product states and findings of cost and schedule for individual capabilities.

Acquisition processes. The process should support sprints and releases. Developmental testing is incremental and continuous. Operational testing needs to leverage developmental testing, accommodate modular testing of incremental changes, and reflect operational capabilities as viewed by users.

Program structures. Programs should be structured to support the degree of "Agile" development identified in the strategy and should integrate users into the acquisition and iteration process to capitalize on feedback for future development. In highly "Agile" development programs, users should be accessible to development and contracting specialists.

Headquarters structures. Unaffected, although strategic support and workforce training and empowerment are needed. Also, if "Agile" development is new to an organization, then some leadership support and guidance might be needed.

Requirements. Requirements need to be flexible, reflecting broad general intent. They should be stated in terms of the mission to be enabled and the minimum levels of capability that are valuable. Specific details of system capabilities are determined interactively through user feedback on incremental builds.

Budgeting. Budgeting can remain fixed if it contains the whole effort with multiple spins, delivering the best capabilities possible within set funding.

Downsides and trade-offs. As with incremental acquisition, "Agile" development might take longer to deliver the full capabilities desired (if the acquisition proceeds to that point). It also requires periodic user feedback and an ability to adjust requirements during acquisition.

General

Descriptions

In-house engineering and production. This approach involves activities in which the government performs engineering and production at government facilities under government control. They might use contractors for some functions. This approach can be faster in that engineering and production capabilities are already on board, avoiding new source selections. Also, in-house capabilities might be faster in that any needed changes can be made without having to renegotiate the engineering or production contracts with an external contractor.

Formal, active lessons learned processes. This is an explicit activity in which the producer and the government have processes in place to continuously evaluate, document, improve, and observe the effects of improvements in a procurement process. The focus is on sharing ideas for more-responsive or faster acquisition processes. The lessons learned process is an enabler, not a responsive approach in and of itself.

Standing FAR waivers. Under appropriate conditions, this approach waives normal acquisition regulations as a way to facilitate a more streamlined (faster, cheaper) administrative process. This approach might be faster depending on what default acquisition process is waived. For example, waiving delivery of documents for a rapid acquisition might enable an effort to proceed with documents to be written and delivered in parallel.

Conditions for Use

Table 2.3 lists the key conditions generally necessary for use of each approach and a short discussion of the reasoning behind the conditions.

21

Table 2.3. Key Conditions: General

Approach	Necessary Condition(s)[a]	Reasoning
In-house engineering and production	• Highly skilled government staff • Government staff capacity • Government development	In-house engineering and production (organic government development) requires the existence of organic government capacity and expertise to develop a capability. This approach contrasts with the usual use of prime contractors to develop a capability.
Formal, active lessons learned process	• Formal feedback loop for training or education	Learning from lessons can be valuable in any situation.
Standing FAR waivers	• FAR exemptions consistently applicable	If an exemption applies consistently for statutory or regulatory conditions (e.g., for contracting), then a standing waiver could be made ready to speed application.

[a] Generally, each approach needs all the features listed (not just one).

Implementation Considerations

In-House Engineering and Production

Acquisition strategies. This strategy relies on using flexible in-house engineering staff rather than contracting out for products or services throughout the course of the program.

Acquisition processes. The contracting process focuses on maintaining flexible in-house capabilities. Review processes focus on overseeing and ensuring efficient management of in-house capabilities rather than of work by prime contractors.

Program structures. Programs are structured to manage and leverage in-house capabilities instead of managing and overseeing external contracted work by prime contractors.

Headquarters structures. Unaffected, but some management of in-house engineering might be needed.

Requirements. Unaffected.

Budgeting. Program-level budgeting is unaffected, but spending within the program is more on in-house support and less on prime contractors.

Downsides and trade-offs. Having an in-house capability implies that it must be created and maintained rather than outsourced. That requires longer-term strategic planning and budgeting to maintain and manage as workloads rise and fall. The use of support contractors can improve some flexibility in capacity and skills compared with staffing by government employees, but even outsourced workforces need some planning and stability.

Formal, Active Lessons Learned Process

Acquisition strategies. This strategy reflects comprehensive exploration of alternatives rather than provision of a narrow, predefined solution space. It also reflects ongoing consideration of alternatives.

Acquisition processes. This is described in the strategy section, above.

Program structures. Programs are structured for flexible, ongoing consideration of alternatives (e.g., in requirements management, contracting, and budget allocation).

Headquarters structures. Unaffected beyond enabling and supporting broad, active analyses of alternatives.

Requirements. Requirements need to be ongoing, with flexible consideration of alternatives.

Budgeting. There should be flexible program-level allocation of resources as alternatives are selected.

Downsides and trade-offs. A system to preserve and share lessons is often a good idea, but contributing to such a system (let alone establishing and managing one) can be a drain on time and on financial and personnel resources. Access questions can also be a consideration: too narrow, and the benefits will be limited; too broad, and security concerns might ensue.

Standing FAR Waivers

Acquisition strategies. This strategy requires thinking through which FAR regulations should be waived and why. The FAR codifies lessons and best practices, but there might be strategic reasons in some cases to waive regulations as long as laws are followed. If waivers are available by default (e.g., MTA), the strategy needs to ensure that the entrance criteria are appropriate and do not cause unintended negative side effects (e.g., whether a rapid prototype or rapid fielding can be accomplished within the five-year time frame).

Acquisition processes. Processes should provide streamlined ways to justify and approve waivers, even in contexts in which PMs are encouraged to highly tailor their acquisition strategies. The acquisition process associated with waived FAR regulations needs to be able to be readily bypassed without staff interference. For default waivers (e.g., such as the JCIDS waiver in MTA), processes need to be put in place so that review staff understand when a program is in a waiver track and avoid adding restrictions back in.

Program structures. Any effects on program structures depend on what regulations are waived. For example, if an OT is focused on testing to user acceptance rather than a fixed set of testing requirements, then the program's testing structure might need to implement direct user involvement in tests to speed feedback and understanding of the status of the system under development. Some default process waivers (e.g., MTA) can drive program structures to ensure that they satisfy the system's use criteria (e.g., the program must facilitate testing and deployment within a prescribed time limit).

Headquarters structures. Oversight needs to be structured so that it encourages tailoring and quick review and approval of the application of waivers. In cases of default waivers (e.g., in MTA), headquarters needs to ensure proper guidance and training for PMs and staff to understand the appropriate use of the waivers.

Requirements. These depend on whether waivers affect the requirements process and how (e.g., whether JCIDS is bypassed).

Budgeting. Any budgeting effects depend on what is waived, but the Anti-Deficiency Act remains in force.

Downsides and trade-offs. Standing waivers might be limited by applicability (i.e., if they can be a default, then why does the process being waived exist?) or lead to a risk of abuse (i.e., there are often reasons behind the waived process, so having a default waiver might disregard those considerations too much).

Intelligence-Related

Descriptions

Faster intelligence. The ability to build capability to the latest and emerging threats is critical to developing relevant warfighter capabilities. This approach seeks to transmit needed intelligence more quickly to the acquisition effort (e.g., through such techniques as the Validated Online Lifecycle Threat [VOLT] reports). Although changes in response to the new threat information might actually slow programs down relative to their original schedules, shortening the time it takes for programs to get needed intelligence will save time for programs that must stay current against threats. In other words, this might minimize schedule growth while enabling agility and responsiveness to threats.

Conditions for Use

Table 2.4 lists the key conditions generally necessary for use of each approach and a short discussion of the reasoning behind the conditions.

Table 2.4. Key Conditions: Intelligence-Related

Approach	Necessary Condition(s)	Reasoning
Faster intelligence	• Threats are changing	When programs must adjust to changing threats, getting intelligence faster on those threats will speed that responsiveness. Of course, not making changes at all is even faster, but that is a different approach and might not be what is deemed necessary by DoD leadership and Congress for the program.

Implementation Considerations

Faster Intelligence on Threat Changes

Acquisition strategies. This reflects the strategy and approach for seeking and managing early and ongoing intelligence on threat changes and the approach for seeking capabilities to address new threats.[7]

---

[7] Generally, making program modifications based on intelligence to meet changing threats will slow programs. However, if such changes are unavoidable or necessary, processes that get the intelligence to the program faster will shorten the response time to those threats.

Acquisition processes. This approach makes explicit use of processes for seeking, managing, and responding to intelligence (e.g., VOLT assessments). Other acquisition processes should remain unaffected except for handling faster responses to threat changes. Although this approach helps ensure that relevant capabilities reach the field, it can slow down programs because of scope creep.

Program structures. Program structures will need (1) an explicit function to utilize these intelligence challenges and (2) associated management of intelligence and of any program and requirement changes that result from changing threats.

Headquarters structures. Headquarters structures are generally unaffected, but leadership would need to be ready to deal with any major programmatic redirections, requirement changes, or budgetary gaps resulting from changing threats.

Requirements. Faster intelligence on threats might result in a need to flexibly and quickly adjust requirements to address new threats.

Budgeting. Faster intelligence on threats might result in a need to flexibly and quickly adjust budgetary resources to address new threats. Ideally, however, the original budget would allow for resources to accommodate changing threats (i.e., the budget should anticipate that threats might change and that the program will need some resource pools to draw on).

Downsides and trade-offs. Establishing and managing intelligence-sharing systems can consume time, as well as financial and personnel resources. Access questions can also be a consideration: too narrow, and the benefits will be limited; too broad, and security concerns might ensue.

Requirements-Related

Descriptions

Requirements: Partial (80%) solutions acceptable to users. Increasingly, agencies and programs are willing to accept a partial solution to a problem (e.g., something that does most of the things desired but not the things that would delay acquiring the other capabilities) if it can be delivered faster (or cheaper). A willingness to accept an iterative or partial solution to have acceptable capabilities sooner means having the ability to choose which requirements to meet and which to delay or forgo to meet operational demands (National Defense Industrial Association [NDIA], 2012; NDIA, 2016; GAO, 2015). An example of this strategy being used successfully is the development of the P-8A Poseidon maritime patrol aircraft. During development, requirements were scaled back to meet schedule, with the expectation that future blocks would integrate the delayed capabilities. Although the initial batch of P-8s were in some ways less capable than their predecessor (the P-3C), the P-8A program is generally regarded as one of the more successful Acquisition Category (ACAT) I programs in the modern era.

Requirements: Keep stable (avoid creep). Shifting requirements can add cost and delays to programs. One way to prevent this is to reach an agreement on an acquisition strategy with the requirements community at the beginning of a project or program that limits the amount of scope creep or requirements shift that can occur during the development of the first increment of the system. Here, the focus is on getting initial capabilities to the field and dealing with threat changes or technical opportunities later in modifications and upgrades. This can result in capabilities that are less operationally relevant if the product has not been adapted to the latest threats.

CSBs. These boards are established by Component Acquisition Executives (CAEs) "to review all requirements and significant technical configuration changes that have potential to affect cost and schedule of Acquisition Category (ACAT) I and IA programs" and make adjustments as necessary (DoDI 5000.2). Thus, CSBs can be a forcing function to stave off delays from requirement increases or to facilitate reductions in requirements as schedule implications arise during program execution. Although CSBs are established in policy for Major Defense Acquisition Programs (MDAPs) (DoDI 5000.02; 10 USC §2430), we included them for longitudinal completeness because the concept was established explicitly to facilitate trade-offs during acquisition to speed programs and control costs. Also, the use of CSBs might be useful for smaller acquisition programs.

Bypass JCIDS and JROC. This approach describes programs in which the government (program) does not have to follow the JCIDS process or undergo review by the JROC as established in 2018 by Chairman of the Joint Chiefs of Staff Instruction (CJCSI) 5123.01H. Examples are acquisitions under the statutory waiver authority in MTA (10 USC §2302; also, see the discussion below) or requirements for a single military service that are not designated as "joint performance requirements" by the Joint Staff Gatekeeper (10 USC §181; CJCSI 5123.01H). This is intended to reduce the amount of administrative overhead from added Joint Staff requirements validation associated with procuring a new technology, thus speeding it to production.

Conditions for Use

Table 2.5 lists the key conditions generally necessary for use of each approach and a short discussion of the reasoning behind the conditions.

Table 2.5. Key Conditions: Requirements-Related

Approach	Necessary Condition(s)[a]	Reasoning
Requirements: Partial (80%) solutions acceptable to user	• Flexible acceptance levels regarding requirements	Recipients are willing to consider partial satisfaction of requirements with prioritization based on feedback regarding cost and schedule implications during development. Explicit strategies to reduce risks from requirements that turn out to be more difficult, time consuming, or costly than originally thought often hinge on the ability to trade requirements during development as these facts are uncovered.
Requirements: Keep stable (avoid creep)	• Incremental capabilities useful	If incremental capabilities are useful, then avoiding requirements creep (e.g., dealing with evolving threats) can speed initial capability delivery.
CSBs	• Requirements might be tradable depending on cost, schedule, or other issues that emerge during acquisition	CSBs provide leadership approval for requirements changes using feedback regarding cost and schedule implications during development.
Bypass JCIDS and JROC	• Joint issues largely absent • Requirements approval authority at the Component level	A key reason why the JCIDS process with JROC approval was established was to ensure that joint issues and aspects are considered in the design of new capabilities. Thus, the JCIDS and JROC process should be bypassed only when joint issues are largely absent and the Component has the authority to approve the requirement.

[a] Generally, each approach needs all the features listed (not just one).

Implementation Considerations

Requirements: Partial (80%) Solutions Acceptable to Users

Acquisition strategies. This strategy reflects an explicit approach of identifying a solution that fits most (but not necessarily all) specified requirements, thus implementing a solution that satisfies the user in a timely (and probably more cost-effective) fashion. Feedback or insight from the user community is a key element.

Acquisition processes. Processes are adjusted to pursue high-value (in schedule) requirements rather than meeting all requirements. CSBs can help in adjusting requirements.

Program structures. Programs are structured to focus requirements management on determining which requirements are high in value rather than meeting all requirements. Program structures also reflect the need to obtain any approvals for changes in requirements.

Headquarters structures. Unaffected, although leadership support is important, structures such as CSBs can be useful, and an explicit process or authority approach is needed to achieve requirements flexibility.

Requirements. This approach directly affects how programs manage and meet acceptable levels of requirements. One necessity is user willingness to forgo requirements that are not absolutely essential and to accept partial capabilities.

Budgeting. Basic budgeting processes are unaffected, although this approach might lead to fewer rebudgeting issues (i.e., budget stability) because requirements relief is possible.

Downsides and trade-offs. There might be concerns that the resulting solutions might become too focused on near-term rather than far-term needs, so active consideration of time frames might be needed. Also, this approach is generally thought to be desirable but often overlooked because of existing requirements processes in such a huge organization as the DoD. This approach requires access and interactions with the user community to share insights learned during development and production, including cost and schedule implications for various capability requirements.

Requirements: Keep Stable (Avoid Creep)

Acquisition strategies. Keeping requirements stable for the first (current) increment is a strategy focused on delivering initial capabilities faster and dealing with threats or technological opportunities in subsequent upgrades. The strategy needs the support of the requirements community and leadership.

Acquisition processes. Unaffected, although CSBs can help minimize requirements changes.

Program structures. Unaffected.

Headquarters structures. Unaffected, although leadership support is important and such structures as CSBs can be useful.

Requirements. Discipline and a willingness to accept incremental capabilities is important.

Budgeting. Unaffected, although requirements stability helps keep budgeting stable.

Downsides and trade-offs. Stable (fixed) requirements might not be responsive to evolving threats.

Configuration Steering Boards

Acquisition strategies. This strategy is based on an acknowledgement of the existence of a governing CSB and its potential for use when seeking requirements changes as technical or budgetary problems arise or threats change.[8]

Acquisition processes. CSBs streamline processes for changing requirements during program execution.

Program structures. As with satisficing requirements, programs are encouraged to raise concerns about challenging requirements rather than seeking to meet absolutely all requirements regardless of cost and schedule.

---

[8] Although CSBs are established in policy for MDAPs (DoDI 5000.02; 10 USC §2430), we included them for longitudinal completeness because the concept was established explicitly to facilitate trade-offs during acquisition to speed programs and control costs. Also, the use of CSBs might be useful for smaller acquisition programs.

Headquarters structures. A CSB must be established, replacing other headquarters structures related to requirements processes.

Requirements. The ways in which programs manage and meet requirements are directly affected.

Budgeting. Basic budgeting processes are unaffected, except that this approach might lead to fewer rebudgeting issues because requirements relief is possible.

Downsides and trade-offs. CSBs might be limited by the amount of empowered senior leadership time available.

Bypass JCIDS and JROC

Acquisition strategies. The acquisition strategies need to reflect who the final requirement authority is and how any flexibility or adjustments might be made. Ideally, the strategy would also have to take into account whether the program should be joint and why (or why not) given that the joint requirements process is being bypassed.

Acquisition processes. The replacement requirements process needs to be clarified in the acquisition strategies (see above).

Program structures. Portions of the program office that deal with managing requirements need to be modified accordingly.

Headquarters structures. Review and management processes for headquarters requirements will need to be clarified or established.

Requirements. See above.

Budgeting. If requirements are expected to change in response to changing threats or technical opportunities, flexible financial resources would be needed.

Downsides and trade-offs. Bypassing the Joint Staff might result in insufficient consideration of joint (versus single-service) considerations important for joint DoD operations.

Schedule-Based

Descriptions

Crashing the schedule. This approach involves analyzing the paths in the schedule, identifying which ones are affecting the overall schedule, assessing ways in which they could be accelerated by additional investments, and optimizing based on available resources (for example, see Defense Systems Management College [DSMC], 2001; Mantel et al., 2011; NDIA, 2012; NDIA, 2016; and GAO, 2015). Examples are investing in lead R&D items to increase maturity and reduce subsequent work; increasing spending on critical tasks; making calendar adjustments, such as adding overtime, adding shifts, or reducing vacation; and pressing exempt salaried employees to work longer hours. In addition to requiring additional resources, crashing might affect quality, involve performance compromises, or necessitate requirements relief or adjustments. Also, simply applying more labor on a task does not necessarily mean that it can be accelerated (i.e., the "mythical man-month"[9] is still in play; see Brooks, 1975; and Brooks, 1995).

Fast-tracking (parallelization; concurrency). This approach identifies tasks that can be broken into pieces that can be executed in tandem (possibly with adjustments to their interdependencies) or with partial overlaps (leads) to accomplish the overall task in less time (Mantel et al., 2011; NDIA, 2012; NDIA, 2016; GAO, 2015). This approach might introduce additional work to deal with any remaining interdependencies; it might also involve overlap (concurrency) between the design and production phases (Mantel et al., 2011). Caution is warranted in having too much concurrency because changes in designs might need to be reflected in items already produced (for example, problems with high concurrency in the F-35 program; see Clark, 2012; and Kendall, 2013).

Streamlining. This approach seeks alternative (faster) ways to complete the tasks and meet the requirements (NDIA, 2012; NDIA, 2016). It might involve reusing parts of work completed elsewhere, innovating faster methods, or identifying low- or non-value-added work that is not critical for meeting key requirements. Explicit consideration and management of risks might be warranted to avoid pushing the schedule so fast that the program fails (i.e., "run to fail").

Focused work. In this approach, the PM identifies critical program tasks and takes an active role to ensure successful completion of these tasks (DSMC, 2001; NDIA, 2012; NDIA, 2016; GAO, 2015). Steps might include reducing multitasking and barriers on critical paths or leveraging slack time on noncritical tasks to float those efforts and focus work on the critical paths. This differs from crashing the schedule in that efforts are made to adapt and "protect the critical/driving path" (NDIA, 2012; NDIA, 2016) rather than investing in ways to speed those critical paths. This approach involves a program culture of staff support.

Work-breakdown schedule reviews. PMs review the relationships and basis of the planned schedule to identify any adjustments that could shorten delivery time. This approach reviews constraints between tasks, lead and lag relationships, the basis of lag assumptions, and duration estimates (DSMC, 2001; NDIA, 2012; NDIA, 2016; GAO, 2015).

Schedule reserve (margin; contingency). Analogous to a budgetary MR, a schedule reserve (i.e., a time reserve, margin, or schedule contingency) gives the PM a tool for more efficiently managing risks and unforeseen issues (DSMC, 2001; NDIA, 2012; NDIA, 2016; GAO, 2015). Although the addition of a schedule margin might lengthen the initial schedule, it is intended to introduce flexibility to deal with problems that arise and avoid major ripples throughout the effort that would result in even more-significant delays. As with budgetary reserves, caution is needed to ensure that staff, contractors, or subcontractors do not view the reserve as an excuse for not trying to meet the original schedule.

---

[9] Brooks argues that *"the man-month as a unit for measuring the size of a job is a dangerous and deceptive myth. . . . Men and months are interchangeable commodities only when a task can be partitioned among many workers with no communication among them"* (Brooks, 1995, p. 16). Thus, simply adding more labor may not make the effort go faster.

Readiness-based gates or milestones. This is a more traditional schedule-based approach to proceed with development. This approach, based on meeting milestones for performance in tests, is more suitable for advanced technologies for which development and roadblocks are less certain. Condition-based and readiness-based reviews can minimize schedules (cycle times) by advancing programs when they are ready rather than delaying until the original review dates for advancement. Of course, if a program's work is showing schedule growth, then those delays would result in a delay of passing milestones, so this approach focuses more on enabling programs that can go faster so that the acquisition process is not slowing it down.

Hard, fixed schedules. This approach, which emphasizes meeting specific goals at specific times for a project, allows for greater contractor accountability and is likely to be more suitable for COTS, GOTS, or mature technology integration programs. Fixed schedules can address the flip side of the condition-based and readiness-based reviews because they can drive a program that is lagging to make trade-offs and stay on schedule rather than allowing schedule growth.

Conditions for Use

Table 2.6 lists the key conditions generally necessary for use of each approach and a short discussion of the reasoning behind the conditions.

Table 2.6. Key Conditions: Schedule-Based

Approach	Necessary Condition(s)	Reasoning
Crashing the schedule	• Budgetary resources obtainable	Crashing the schedule involves making investments to shorten tasks on the critical schedule path (DSMC, 2001; Mantel et al., 2011; NDIA, 2012; NDIA, 2016; GAO, 2015).
Fast-tracking (parallelization; concurrency)	• None (universal)	If opportunities exist to parallelize tasks or introduce concurrency, then this approach is universally applicable. Caution is warranted, however, because risk might increase.
Streamlining	• None (universal)	One can always try to look for innovative, faster methods, reuse work completed elsewhere, or eliminate low-value work.
Focused work	• None (universal)	Taking an active role in supporting and protecting critical tasks should be something any PM can undertake.
Work-breakdown schedule reviews	• None (universal)	Review of the work-breakdown schedule is a common approach in program management.
Schedule reserve (margin; contingency)	• None (universal)	Schedule reserves can be built into any program schedule.

Approach	Necessary Condition(s)	Reasoning
Readiness-based gates or milestones	• None (universal)	Defining approval gates or milestones based on the condition of the program at key points in the schedule is applicable to any program, especially if these gates and milestones are tailored to each program (e.g., having a development maturity gate only for developmental programs). The scheduling of these reviews should occur when the program is ready for the review, not on a fixed schedule.
Hard, fixed schedules	• Flexible acceptance levels for requirements	Setting inflexible schedules can be considered when requirements are tradable for schedule. Note also that fixed schedules might also lead to cost growth when extra resources must be applied to deal with problems that arise.

Implementation Considerations

Crashing the Schedule

Acquisition strategies. The strategy should mention that schedule is a priority and identify the resources and support available to invest in shortening critical tasks.

Acquisition processes. Unaffected.

Program structures. The program is structured to identify critical paths, assess investment options, and make changes and investments accordingly.

Headquarters structures. Unaffected.

Requirements. Unaffected.

Budgeting. Extra funds should be made available to invest in shortening the schedule (DSMC, 2001; Mantel et al., 2011; NDIA, 2012; NDIA, 2016; GAO, 2015).

Downsides and trade-offs. This is a common practice but does require PM attention and financial resources.

Fast-Tracking (Parallelization; Concurrency)

Acquisition strategies. The strategy should identify key elements, the level of concurrency incurred, and any associated risks so that leadership can concur on the approach.

Acquisition processes. If significant concurrency between development and production is envisioned, then the oversight reviews and schedule will need to be planned accordingly.

Program structures. The program needs elements to identify fast-track opportunities, implement them, and manage risks and any new interdependencies.

Headquarters structures. Unaffected.

Requirements. Unaffected.

Budgeting. Unaffected.

Downsides and trade-offs. Tightened schedules can increase the risk that problems that arise will ripple quickly through the program and result in frequent or significant restructuring. Significant concurrency between design and production can result in significant risks of cost and schedule growth and performance limitations (e.g., the issues with the F-35 concurrency mentioned earlier—see Clark, 2012; and Kendall, 2013).

Streamlining

Acquisition strategies. The strategy should identify which streamlining methods (e.g., reuse, innovation, work optimization) will be considered, along with any associated risks for oversight cognizance and approval.

Acquisition processes. Any effects on the process depend on the streamlining methods used.

Program structures. The program management structure and processes must directly implement this approach.

Headquarters structures. Unaffected.

Requirements. Unaffected.

Budgeting. Some innovative approaches might be more expensive.

Downsides and trade-offs. This is a common practice but does require PM attention and resources.

Focus Work

Acquisition strategies. Beyond specifying a cultural shift to focus work on the critical path, specific barriers or processes known in advance might be mentioned in the strategy.

Acquisition processes. Processes that affect key tasks might need adjusting (e.g., establishing integrated product teams to parallelize multifunctional reviews, integrating developmental and operational testing, tailoring or streamlining document coordination).

Program structures. The program management structure and processes must directly implement this approach.

Headquarters structures. Acquisition process changes might involve headquarters structures.

Requirements. Unaffected.

Budgeting. Unaffected.

Downsides and trade-offs. This is a common practice but requires both PM attention and support from the larger organization and its leadership.

Work-Breakdown Schedule Reviews

Acquisition strategies. The strategy should feature a proactive program management structure in which the PM is involved in examining the work-breakdown structure of contracted efforts.

Acquisition processes. Unaffected.

Program structures. Unaffected.

Headquarters structures. Unaffected.

Requirements. Unaffected.

Budgeting. Unaffected.

Downsides and trade-offs. This is a common practice but does require PM attention.

Schedule Reserve (Margin; Contingency)

Acquisition strategies. Use of a schedule reserve might be explicitly mentioned in the strategy.

Acquisition processes. Unaffected.

Program structures. The program structure and processes need elements to explicitly manage the schedule reserve to ensure that staff, contractors, or subcontractors do not view the reserve as an excuse for not trying to meet the original schedule.

Headquarters structures. Unaffected, although (as with budgetary MRs) leadership must protect the schedule reserve if it is to be an effective tool for the PM.

Requirements. Unaffected.

Budgeting. Unaffected.

Downsides and trade-offs. As discussed earlier, such reserves can lengthen the initial schedule and risks abuse by staff or contractors. If visible, reserves need support of leadership.

Readiness-Based Gates or Milestones

Acquisition strategies. This strategy features ongoing status assessments and drivers to push schedules and seek passage of gates or milestones.

Acquisition processes. The review process (gates or milestones) needs to be flexible in scheduling programs when they are ready.

Program structures. Programs need (potentially more) explicit self-evaluations of readiness for reviews.

Headquarters structures. Basic structures might be the same, although headquarters review of oversight entrance criteria might be affected.

Requirements. Unaffected.

Budgeting. Unaffected, although shifts in reviews (earlier or later than expected) might involve budgetary considerations (e.g., budgetary decreases or increases).

Downsides and trade-offs. As discussed earlier, using conditions to set milestones and gates can remove those reviews as forcing functions to pressure schedules and might result in schedule growth.

Hard, Fixed Schedules

Acquisition strategies. This strategy features approaches for how to stay on schedule as difficulties arise (e.g., seeking requirements relief, having alternative technical developments for high-risk areas, trading cost growth to keep on schedule)

Acquisition processes. Unaffected, except that schedules are recognized as being fixed.

Program structures. Unaffected, although program structures must focus on meeting schedules—perhaps at the expense of cost or requirements.

Headquarters structures. Unaffected.

Requirements. Fixed schedules might result in needs to reduce requirements.

Budgeting. Fixed schedules might result in increased costs.

Downsides and trade-offs. Fixed schedules might result in cost growth, failure to meet requirements, or lack of responsiveness to changing threats or important technical opportunities.

Financial-Related

Descriptions

DoD Rapid Prototyping Fund. This account allows for funds to be set aside specifically for rapid development of technologies to the level of working prototypes. This approach allows a high degree of freedom for PMs to make project-funding decisions based on emerging threats or technology opportunities outside the traditional Program Objective Memorandum (POM) cycle. The ready availability of funds could speed a program and make it more responsive to needs rather than waiting for new budgets to be passed or for the financial community to identify funds that can be reprogrammed for the prototype.

Reprogramming. This describes the ability of the DoD or government to move funds around among projects and programs more flexibly to meet emergent requirements (10 USC §2214). Such flexibility could facilitate responsiveness and fund initial work while efforts to secure longer-term funding are pursued.

Budget stability measures. Programs that have relatively steady and reliable funding can operate in a more methodical, planned way. Budget uncertainty results in programmatic decisions that emphasize the short term and can result in unnecessary administrative and programmatic churn. Stability can avoid (1) delays from unplanned ramp-ups in production (when budgets are increased over plans) and (2) lengthening production time (when budgets are cut and production must then be stretched in time). Some potential techniques would be leadership placing explicit strategic importance on holding budgets stable, leadership being provided with an upfront acquisition strategy that lays out the need for budget stability in particular phases of the acquisition, or leadership having a clear understanding of the implications on an acquisition of budget instabilities. Budget stability is particularly challenging: Authorities are not localized within a military service or even within the DoD because Congress ultimately holds the purse strings. Additional ideas as they might pertain to S&T budgets are discussed in Appendix C.

MRs. This approach relies on funding not allocated for any specific program or project that the PM has authorization to use as needed on existing programs to meet emergent, shifting, or expedited requirements (e.g., an adversary acquires an asymmetric capability far sooner than expected). Reserves can be used to address these new needs in a more responsive fashion or to invest selectively in schedule mitigations when problems occur with tasks on the critical path. This option can also be a part of a PM's risk mitigation plans and strategy.

Conditions for Use

Table 2.7 lists the key conditions generally necessary for use of each approach and a short discussion of the reasoning behind the conditions.

Table 2.7. Key Conditions: Financial-Related

Approach	Necessary Condition(s)[a]	Reasoning
DoD Rapid Prototyping Fund (Section 804; 10 USC §2302 Historical and Revision Notes)	• Prototyping possible in short time • Technology relatively mature • Funding available in prototyping account	The Rapid Prototyping Fund is applicable when prototyping can be accomplished in the statutory allotted time (i.e., "field a prototype that can be demonstrated in an operational environment and provide for a residual operational capability within five years of the development of an approved requirement" [Section 804; 10 USC §2302 Historical and Revision Notes]). The time constraints usually imply that the technology being applied is relatively mature because little time is available to further develop and integrate the technology.
Reprogramming	• Budgetary resources obtainable • Budget needs relatively small in near term • Ability to influence budgets	Reprogramming can be used when the budget needs are relatively small in the near term and thus fit within the statutory limits.
Budget stability measures	• Ability to influence budgets	Budget stability measures are useful generally but require some ability to influence budgets. Stability measures might be needed (but are harder to enforce) for acquisitions below the top-priority programs because those programs tend to be the targets of budget cuts when resources are dropping or urgent needs arise.
MRs	• None (universal)	Having an MR to deal with unforeseen problems that arise in a program is a best practice that is universally applicable. The challenge is not one of applicability but one of whether the operating environment and leadership for a program will allow and protect the MRs for program use.

[a] Generally, each approach needs all the features listed (not just one).

Implementation Considerations

DoD Rapid Prototyping Fund

Acquisition strategies. This strategy features prototyping and the use of the DoD Rapid Prototyping Fund.

Acquisition processes. Unaffected, except that procedures need to be in place for using the fund.

Program structures. Unaffected, except that this applies only to programs using prototyping activities.

Headquarters structures. Unaffected, except that procedures need to be in place for using the fund.

Requirements. Unaffected.

Budgeting. Use of the fund can ease budgeting at early (prototyping) stages and thus speed program initiation.

Downsides and trade-offs. As mentioned, money must be in the fund for it to be useful.

Reprogramming

Acquisition strategies. This strategy reflects (1) whether reprogramming might be sought to fund program initiation or perhaps (2) how the program might respond if reprogramming marks reduce this program's budget (especially if the program is known to have relatively low priority or might be viewed as a bill-payer commodity program).

Acquisition processes. Unaffected.

Program structures. Unaffected.

Headquarters structures. Unaffected.

Requirements. Unaffected, unless the program loses resources and must seek requirements relief.

Budgeting. Unaffected.

Downsides and trade-offs. Reprogramming is limited, can consume time and labor, and can require uncertain support by Congress. Also, the money comes from somewhere; therefore, there can be significant negative consequences on the program or organization from which the funds are taken (depending on the situation).

Budget Stability Measures

Acquisition strategies. If the program is a high priority with strong leadership support, then the strategy can focus on elements other than approaches for coping with unstable budgets.

Acquisition processes. Unaffected.

Program structures. Unaffected, although budget stability can make program execution easier and less stressful.

Headquarters structures. Unaffected.

Requirements. Unaffected.

Budgeting. Stable budgets are largely a function of the POM process, program priority, and leadership support. Some programs in production are better able to handle annual changes in budgets, but caution is warranted even in these cases because changes in quantity can lead to inefficient production as companies adjust capacity and cannot make multiyear investments or purchases of subcomponents.

Downsides and trade-offs. As mentioned, stability requires leadership support (even when leaders change). Also, on the flip side of reprogramming, budget stability can limit the ability to obtain money in the short term when new critical needs arise. Longer-term budget stability also requires not only leadership support but also support across multiple sessions of Congress that have uncertain, changing political positions.

Management Reserves (Financial)

Acquisition strategies. This strategy should explicitly reflect that an MR exists, how large it should be, some consideration of what it might be used for (generally), and that it needs leadership support to protect the reserves during budget reviews.

Acquisition processes. Unaffected, although programs should be more stable as a result, requiring less intervention.

Program structures. Programs have an explicit MR in financial-management structures.

Headquarters structures. Explicit leadership support and defense of MRs are probably needed, including protection of MRs when in-year reprogramming searches are conducted to fund critical needs.

Requirements. Unaffected, although MRs lower the risk of meeting requirements.

Budgeting. Explicit leadership support and defense of MRs is probably needed, including protection of MRs when in-year reprogramming searches (or execution reviews) are conducted to fund critical needs. MRs should lead to more-stable program budgets (i.e., programs should be able to handle problems better within existing budgets).

Downsides and trade-offs. Financial MRs require leadership support to survive the frequent reviews that search for money to reprogram (and PM time and effort to defend the MRs).

Technology Development–Related

Descriptions

Operational prototyping: new platforms. This refers to "from scratch" prototypes that are not developments of existing platforms. Operational prototypes can be faster (i.e., capabilities are available to operational users during the prototyping phase—albeit with performance and reliability risks). Prototypes can be large or small. Examples of large, high dollar–value prototypes would be DARPA's Sea Hunter and the Global Hawk. These might become programs of record, as did Global Hawk. Smaller dollar–value projects could include development of small unmanned aerial vehicle swarms. Such programs would not result in more than an ACAT III program, even if brought to full production.

Operational prototyping: MTA rapid prototyping (Section 804). MTA is a specific type of rapid acquisition approach that focuses on delivering capability in a period of two to five years. Interim guidance was issued by the Under Secretary of Defense for Acquisition and Sustainment (USD[A&S]) in fulfillment of Section 804 (which describes the "middle tier of acquisition for rapid prototyping and rapid fielding") (Pub. L. 114-92, 2015; 10 USC §2302 Historical and Revision Notes). MTAs are not subject to JCIDS, DoD Directive (DoDD) 5000.01, and associated instructions (e.g., DoDI 5000.02). The objectives of MTA prototyping are to field a prototype that can be demonstrated in an operational environment and provide for residual operational capability within five years of an approved requirement.

Operational prototyping: components. Development of prototypes is integrated into an existing system, providing some rapid, responsive capabilities during operational testing and use during evaluation and afterward with any left-behind capability. An example might be a frictionless, brushless cooling fan for F-35 avionics.

Operational prototyping: components or technology (Section 806). This is a specific type of component or technology prototyping. Section 806 (which describes the "development, prototyping, and deployment of weapon system components or technology") of the FY 2017 NDAA (Pub. L. 114-328, 2016), added these new authorities for the development, prototyping, and deployment of major weapon system components and technologies. Prototypes of components or technologies that can be completed within two years, demonstrated in a relevant environment, and cost less than $50 million can use cooperative agreements or OTs. Follow-on production or rapid fielding of successful prototypes executed under this statute might, under certain circumstances, employ a noncompetitive contract or OT. Also, up to $50 million in unused procurement funds from other accounts might be used to begin low-rate initial production of the rapid fielding project, although it is not yet clear whether congressional appropriators approve of this repurposing authority. Other details and approvals are specified in 10 USC §2447a–e. The use of this statute (possibly with OTs) might result in shorter development times.

JCTDs. JCTDs are intended to allow the military to evaluate mature technologies and procure them more quickly and cheaply than through the traditional JCIDS process.[10] JCTDs pursue the rapid prototyping development and demonstration of prototypes within two to four years of the identification of a need. JCTDs "affordably operationalize prototyped technologies that enable warfighters to explore novel concepts and to facilitate informed transition to formal programs of record" (Under Secretary of Defense [Comptroller], 2019). Capabilities provided during (or left behind after) a JCTD can serve operational needs in a responsive fashion. Prior examples are the Global Hawk and the Joint Precision Airdrop System.

Prize competitions. Prize competitions are used to set up a competitive environment to challenge researchers, designers, and developers to create innovative solutions. Enabled by 15 USC §3719, such prizes can be an element of innovation but cannot by themselves substitute for funding that enables participants to develop the S&T in the first place. This might be faster in that researchers, designers, and developers are unconstrained in the approaches they employ. More revolutionary high-risk, high-payoff concepts might be explored without a long-term commitment. Also, competitions can inject a level of excitement, especially when people feel a sense of challenge.

---

[10] JCTDs were called Advanced Concept Technology Demonstrations (ACTDs) until 2006.

Conditions for Use

Table 2.8 lists the key conditions generally necessary for use of each approach and a short discussion of the reasoning behind the conditions.

Table 2.8. Key Conditions: Technology Development–Related

Approach	Necessary Condition(s)[a]	Reasoning
Operational prototypes: new platforms	• Prototyping possible in short time • Technology relatively mature	Operational prototyping makes sense when such prototypes can be developed in relatively short order and when the technologies involved are relatively mature to integrate into a new system or are focused on modifications to existing platforms.
Operational prototyping: MTA rapid prototyping (Section 804)	• Budgetary resources obtainable • Prototyping possible in short time • Technology relatively mature	The prototyping authorities under MTA (10 USC §2302 Historical and Revision Notes; Section 804) apply when such prototypes can be developed rapidly and, therefore, when the technologies involved are relatively mature.
Operational prototyping: components	• Prototyping possible in short time • Technology relatively mature	Operational prototyping for components might make sense when such prototypes can be developed in relatively short order and, therefore, when the technologies involved are relatively mature, when prototyping smaller systems (e.g., weapons) or upgrading larger platforms, and/or when such components are on the critical path.
Operational prototyping: components or technology, (Section 806; 10 USC §2447)	• Budgetary resources obtainable • Prototyping possible in short time • Technology relatively mature	The prototyping authorities under 10 USC §2447a–e apply when such prototypes can be developed rapidly and, therefore, when the technologies involved are relatively mature.
JCTDs	• Users involved in exploring capability options • CONOPS can be explored • Value determination needed	Such demonstrations are intended to allow users to explore new system concepts in an operational setting to determine the concepts' value.
Prize competitions	• Risks of technology high • Technology development needed and significant	Prizes can spur innovation and accelerate development, often on hard, high-risk problems that can be appealing as a challenge to researchers and engineers.

NOTE: CONOPS = concept of operations.

[a] Generally, each approach needs all the features listed (not just one).

Implementation Considerations

Operational Prototyping: New Platforms

These practical considerations apply to all prototypes—platforms (large or small) or components. (Also see the more detailed discussion in Appendix A).

Acquisition strategies. In this strategy, prototypes should feature explicit elements, such as what technologies the prototype is expected to mature and demonstrate; whether the prototype is truly representative of the actual system to be produced; and, if there are differences between the prototype and the final system, how the risks associated with those differences will be managed. The strategy should describe the technology development pathway and technology transition plan with measurable and achievable exit criteria.

Acquisition processes. The prototyping phase needs to be reflected in the process.

Program structures. The program structure needs to provide capacity for prototyping design, management, and operational engagement.

Headquarters structures. Unaffected, except that headquarters and MDA must approve operational engagements, ensuring that safety and negative operational effects are minimal and acceptable.

Requirements. Unaffected. Validated requirements are not needed until an actual procurement of prototype units in significant numbers occurs. Some insight into operational needs, however, can be useful for guiding prototype R&D.

Budgeting. Resources for prototyping need to be listed in budgets. If prototypes are left for operational use, then sustainment funding will need to be identified to deploy, operate, and sustain fieldable prototypes immediately on completion of development and test activities.

Downsides and trade-offs. Prototypes might be infeasible for extremely expensive and complex weapon systems (e.g., aircraft carriers and submarines). Other supporting R&D environments, such as the Skunk Works–like approach, might be needed to rapidly develop and produce experimental aircraft prototypes in an efficient fashion outside the formal program. Also, some care should be taken to ensure that prototypes either demonstrate key attributes of the eventual system or that lessons are learned from the effort, avoiding the production of a prototype just to meet a process requirement. (For example, see issues discussed by Kendall, 2017, p. 86.)

Operational Prototyping: MTA Rapid Prototyping (Section 804)

In addition to the general comments on prototyping above, the following specifics apply to Section 804 MTA rapid prototyping (10 USC §2302 Historical and Revision Notes; USD[A&S], 2018b; SAF/AQ, 2018). Below are highlights of what the Secretary of Defense might provide in the guidance. However, specifics are evolving, so we highly recommend consulting the latest guidance from OSD and the military department in question.

Acquisition strategies. The strategy must reflect the restrictions on Section 804 authorities: that the prototype must be demonstrated in an operational environment and the effort must provide a residual operational capability within five years of the development of an approved requirement.

Acquisition processes. Section 804 programs are exempt from DoDD 5000.01—and thus DoDI 5000.02—except to the extent provided in the guidance from the Secretary of Defense (nominally, from USD[A&S]). Also, the guidance might specify that the PM should report directly to the Service Acquisition Executive (SAE) "without intervening review or approval" (Section 804(c)(4)(B)).

Program structures. The guidance might specify that the PM of a Section 804 program can be "authorized staff positions for a technical staff, such as experts in business management, contracting, auditing, engineering, testing, and logistics, to enable the manager to manage the program without the technical assistance of another organizational unit of an agency to the maximum extent practicable" (Section 804(c)(4)(D)).

Headquarters structures. The statute specifies that the PM of a Section 804 program might be provided a process to expeditiously seek a waiver from Congress from any statutory or regulatory requirement that the PM determines adds little or no value to the management of the program (Section 804(c)(4)(G)). Also, the guidance might specify that the SAE (in coordination with the Defense Acquisition Executive) can be MDA (Section 804(c)(4)(F)).

Requirements. Approval of any requirements are exempt from the JCIDS process.

Budgeting. Funds from the DoD Rapid Prototyping Fund might be used. If prototypes are left for operational use, then sustainment funding will need to be identified to deploy, operate, and sustain fieldable prototypes immediately on completion of development and test activities.

Downsides and trade-offs. In addition to the prototyping downsides discussed earlier for rapid prototyping, the use of MTA authorities comes with the specific cost and schedule limitations and management procedures discussed above. Also, the implementation policies and guidance for MTA are in flux, including what degree of oversight and data tracking are required (compared with, say, prototyping efforts outside MTA). In addition, if MTA prototypes are to extend beyond MTA timelines, then transitions to other acquisition approaches and programs might need to be developed and executed.

Operational Prototyping: Components

See the practical considerations for new platform prototypes; the same implementation considerations apply to components.

Operational Prototyping: Components or Technology (Section 806)

In addition to the general comments on prototyping above, the following specifics apply to the Section 806 (10 USC §2447a-e) prototypes.

Acquisition strategies. The strategy must reflect the restrictions on Section 806: that the prototype must be selected by the SAE, completable within two years, demonstrated in a relevant environment, and cost less than $50 million. The strategy can feature the use of cooperative agreements or OTs on projects. Any follow-on production or rapid fielding of successful prototypes executed under this statute might, under certain circumstances, employ a noncompetitive contract or OT (10 USC §2447d(a)).

Acquisition processes. Same as for operational prototyping.

Program structures. Same as for operational prototyping.

Headquarters structures. Section 806 involves "an oversight board or . . . a similar existing group of senior advisors for managing prototype projects for weapon system components and other technologies and subsystems, including the use of funds for such projects, within the military department concerned" (10 USC §2447b), and the SAE is involved in selecting Section 806 activities through a merit-based selection process that meets the criteria set out in 10 USC §2447b(c)(1).

Requirements. Same as for operational prototyping.

Budgeting. The effort under Section 806 must cost less than $10 million (if approved by the SAE) or $50 million (if approved by the Secretary of the military department). Also, up to $50 million in unused procurement funds from other accounts might be used to begin low-rate initial production of the rapid fielding project, although it is not yet clear whether the appropriators approve of this repurposing authority. Other details and approvals are specified in 10 USC §2447a–e. Note that 10 USC §2447a mandates specific budget displays for Section 806 activities in the defense budget materials. If prototypes are left for operational use, then sustainment funding will need to be identified to deploy, operate, and sustain fieldable prototypes immediately on completion of development and test activities.

Downsides and trade-offs. In addition to the prototyping downsides discussed earlier for rapid prototyping, the use of 10 USC §2447 authorities comes with the specific cost and schedule limitations and management procedures discussed above.

Joint Concept Technology Demonstrators

Acquisition strategies. JCTDs are an element of a strategy in which concepts and technology are demonstrated in more-extensive operational demonstrations to gain feedback on needed changes (e.g., in design, operational concepts, technology), reducing risk.

Acquisition processes. Unaffected, JCTDs can fit within traditional tailored or other processes.

Program structures. Unaffected, except that programs should be identified and structured to receive the JCTD should it be successful.

Headquarters structures. Unaffected.

Requirements. At this stage, requirements are often just expressed as general needs.

Budgeting. Plans should be put in the Future Years Defense Program to fund subsequent development should the JCTD be successful. Ideally, these placeholders are flexible to incorporate lessons from the JCTD and to accommodate this or other JCTDs that are not deemed worth continuing into a program.

Downsides and trade-offs. JCTDs can be major efforts that require the involvement of operational or experimental units, systems, and ranges. Thus, extensive planning and resource alignment might be needed.

Acquisition strategies. Prize competitions can be an element of the strategy to spur "the development of solutions for a particular, well-defined problem," "identify and promote a broad range of ideas and practices that might not otherwise attract attention, facilitating further development of the idea or practice by third parties," "create value during and after the competition by encouraging contestants to change their behavior or develop new skills that might have beneficial effects during and after the competition," or "stimulate innovation that has the potential to advance the mission of the respective agency" (15 USC §3719(c)).

Acquisition processes. The head of a federal agency (e.g., the director of DARPA) is authorized to approve programs to "award prizes competitively to stimulate innovation that has the potential to advance the mission of the respective agency" (15 USC §3719(b)).

Program structures. A prize competition might occur before the formal entry point into the acquisition system (i.e., before the Materiel Development Decision).

Headquarters structures. Unaffected, although the agency head will need to approve the prize competition.

Requirements. Prize competitions seek advances in S&T that should be informed by operational needs (before the requirements stage). Objectives might be "technology push" rather than "needs and requirements pull" that might not take new possibilities into account.

Budgeting. Unaffected as long as funding is available. "Financial support for the design and administration of a prize competition or funds for a cash prize purse, might consist of Federal appropriated funds and funds provided by private sector for-profit and nonprofit entities" (15 USC §3719(m)(1)).

Downsides and trade-offs. Prize competitions focus on a financial reward for the winner rather than providing R&D funds for competing teams. Thus, although prize competitions can motivate innovation in a less constrained, unplanned approach, competitions usually rely on the existence of relative mature components or external financial resources to research and develop needed advances.

Reuse

Descriptions

Modifications to existing platforms. Physical or software changes to systems provide them with capabilities that were not part of their original requirements. An example might be tying the observations of several satellites together to provide sensor fusion, or three-dimensional imaging, even if they had not been intended for these capabilities initially. Modifications and component-level upgrades can be faster than fielding entirely new platforms.

Repurposing. This approach involves meeting requirements by using existing assets for a purpose for which they were not intended but would quickly provide a needed capability without having to develop a new system. As an example, Iridium was never intended as a military satellite communications network but has been used as such because it was purchased by the DoD.

Off-the-shelf: COTS or GOTS. These approaches involve acquiring and employing technologies that are already available either as a commercial product or as a system owned or provided by the government. Because the systems or components are readily available, they deliver capabilities faster than if new systems or components are developed for the purpose. However, taking such an approach might involve performance compromises; these approaches differ from repurposing in that technology is used for the purpose for which it was initially intended. COTS acquisitions might or might not be done through JCIDS and JROC processes.

Adaptation of foreign technology. This approach involves purchasing or leveraging foreign technologies to create a capability. An example of this would be building a Global Positioning System (GPS)–denied environment capability into a drone by allowing it to also obtain navigational positioning and timing from Glonass (Russian), BeiDou (Chinese), and Galileo (European) satellite navigation systems. As with COTS acquisitions, using systems or components that are readily available allows faster delivery of capabilities than does developing new systems or components. However, using foreign systems might involve not only performance compromises but also security concerns.

Conditions for Use

Table 2.9 lists the key conditions generally necessary for use of each approach and a short discussion of the reasoning behind the conditions.

Table 2.9. Key Conditions: Reuse

Approach	Necessary Condition(s)[a]	Reasoning[b]
Modifications to existing platforms	• Able to use existing capabilities	Modifications to existing platforms can be used when such an adaptation provides sufficient capabilities to meet the needs (as opposed to being insufficient and requiring a new system).
Repurposing	• Able to use existing capabilities	Repurposing an existing platform can be done when such an adaptation provides sufficient capabilities to meet the needs (as opposed to being insufficient and requiring a new system).
Off-the-shelf: COTS or GOTS	• Able to use existing capabilities	COTS or GOTS components or systems can be applied when such systems provide sufficient capabilities to meet the needs (as opposed to being insufficient and requiring a new system).
Adaptation of foreign technology	• Able to use existing capabilities • Security risks from foreign content is low	Foreign technology can be applied when such an adaptation provides sufficient capabilities to meet the needs (as opposed to being insufficient and requiring a new system). However, leveraging foreign technology makes sense only when those technologies have low security risks.

[a] Generally, each approach needs all the features listed (not just one).
[b] The reasons behind each of these approaches are tautologies to an extent; nonetheless, the approaches can be effective.

45

Implementation Considerations

Modifications to Existing Platforms

Acquisition strategies. This strategy explicitly pursues modifications of existing platforms to accelerate acquisition.

Acquisition processes. Processes are tailored to reflect modifications (along with other factors, such as technical maturity).

Program structures. Unaffected, unless coordinated with other programs—e.g., the base vehicle program requires additional program capabilities.

Headquarters structures. Unaffected.

Requirements. Requirements might need to be adjusted to reflect what is possible under a modified platform. Requirements processes are generally unaffected.

Budgeting. Unaffected, although budgeting levels should reflect costs and potential savings from a modification approach.

Downsides and trade-offs. As mentioned, modifications can constrain the resulting capabilities (compared with the possibilities inherent in designing and producing a new system). Modifications might also incur any known cost, reliability, and performance issues of the existing system.

Repurposing

Acquisition strategies. This strategy explicitly pursues repurposing of existing capabilities.

Acquisition processes. Processes are tailored to reflect repurposing.

Program structures. Unaffected, unless coordinated with other programs—e.g., the base capability program requires additional program capabilities.

Headquarters structures. Unaffected.

Requirements. Requirements might need to be adjusted to reflect what is possible under repurposing. Requirements processes are generally unaffected.

Budgeting. Unaffected, although budgeting levels should reflect costs and potential savings from a repurposing approach.

Downsides and trade-offs. As mentioned, repurposing can constrain the resulting capabilities (compared with the possibilities inherent in designing and producing a new system). Repurposing also incurs any known cost, reliability, and performance issues of the existing system.

Off-the-Shelf: COTS or GOTS

Acquisition strategies. This strategy explicitly focuses on the integration and use of (mature) COTS or GOTS products. The strategy should feature specific approaches for easing the use of COTS or GOTS in acquisitions (e.g., minimizing any modifications and use of commercial interface standards).

Acquisition processes. Processes might be tailored to reflect the maturity of COTS or GOTS products and the use of common interface standards.

Program structures. Program structures might reflect reduced development and component engineering but potentially increased systems engineering to ensure the COTS or GOTS components sufficiently meet requirements.

Headquarters structures. Unaffected.

Requirements. The requirements processes might be unaffected, but the requirements themselves might need to be somewhat flexible to accept the capabilities of available COTS and GOTS products.

Budgeting. Unaffected, although less research, development, testing, and evaluation might be needed.

Downsides and trade-offs. As mentioned, the use of COTS and GOTS systems or components constrains the resulting capabilities (compared with the possibilities inherent in designing and producing a new system). These approaches might also incur any known cost, reliability, and performance issues of the existing system. COTS systems tend to prioritize quick capabilities at the expense of reliability and security, so those concerns need to be taken into account.

Adaptation of Foreign Technology

Acquisition strategies. These strategies are similar to COTS, except that concerns for product security and reliability will need to be addressed.

Acquisition processes. Processes will be needed to test for security and reliability.

Program structures. Program structures need to reflect the potential for added testing.

Headquarters structures. Unaffected.

Requirements. The requirements processes might be unaffected, but the requirements themselves might need to be somewhat flexible to accept the capabilities of available technology.

Budgeting. Unaffected, although resources to address potential security and reliability concerns might be increased.

Downsides and trade-offs. As with COTS, the adaptation and use of existing foreign systems or components constrain the resulting capabilities (compared with the possibilities inherent in designing and producing a new system). This approach might also incur any cost, reliability, and performance issues of the foreign system, and security can be a major concern.

Maturity-Based Modification

Descriptions

Incremental acquisition. An incremental delivery strategy seeks to field initial capabilities as soon as possible, with subsequent increments or block upgrades to deliver more-challenging or time-consuming capabilities. The 2015 update of DoDI 5000.02 explicitly recognizes incremental approaches for software-intensive systems or elements for incremental acquisition, but block upgrades of hardware-dominant systems illustrate this approach. Usually, incremental acquisition differs from "Agile" or component upgrade approaches in that a final capability end state is known and planned for. Each increment or block upgrade is managed as a separate step toward a final, more capable system, and each often has its own set of requirements, review cycles, certification and accreditation, milestones, and acquisition strategies, although smaller block upgrades can be managed together in a single program step.

Continuous component upgrade as technology matures. As with incremental acquisition, a component-upgrade strategy does not wait for technology to mature before delivering initial capabilities. Here, however, the ultimate capability is not planned for from the beginning but is a continuous delivery model. Upgrades are integrated as technologies mature or new advances come to light rather than as a "building block" approach in which specific increments deliver specific capabilities on a schedule. MOSA, with their open interface standards, facilitate component-level upgrades in a plug-and-play fashion. Depending on the size of the component, these upgrades might be made in sustainment, especially for software upgrades.

Conditions for Use

Table 2.10 lists the key conditions generally necessary for use of each approach and a short discussion of the reasoning behind the conditions.

Table 2.10. Key Conditions: Maturity-Based Modification

Approach	Necessary Condition(s)	Reasoning
Incremental acquisition	• Incremental capabilities useful	Incremental acquisition is appropriate only when those incremental deliverables are operationally useful.
Continuous component upgrade as technology matures	• Incremental capabilities useful	Like incremental acquisition, continuous improvement and upgrades as technology matures are appropriate only when those initial incremental improvements are operationally useful.

Implementation Considerations

Incremental Acquisition

Acquisition strategies. This strategy explicitly reflects faster initial delivery followed by incremental builds and features user feedback, testing, and sufficiency determination.

Acquisition processes. Processes must directly reflect the specific incremental approach used and whether each increment needs its own acquisition timeline and reviews (or whether the increments can be overseen in a single stream with common documentation).

Program structures. Programs are structured around the type of incremental approach development employed, including the relationships with users, developers, and testers.

Headquarters structures. Unaffected, although some leadership support is required if an "Agile" acquisition approach is used (e.g., see Kim et al., forthcoming).

Requirements. Requirements need to be more flexible to allow the system to explore the cost–benefit space as incremental improvements are made and user feedback is obtained on what capabilities are satisfactory. Increments are operationally viable and requirements can be parsed to ensure that an increment-based strategy is feasible.

Budgeting. Unaffected if programs work to stay within budgeted resources.

Downsides and trade-offs. Incremental acquisition can require more time to deliver the full capabilities desired. Also, if each increment is reviewed with its own milestone review process, then added effort is imposed.

Continuous Component Upgrade as Technology Matures

Acquisition strategies. This approach is somewhat similar to incremental acquisition, except that the final capability end state is not planned for. Here, capabilities are improved by component upgrades (preferably through standard interfaces) rather than holding up the delivery of early capabilities while waiting for components to mature.

Acquisition processes. Processes reflect an ongoing development and upgrade approach (and associated oversight) rather than the development and delivery of a single configuration.

Program structures. Programs are structured for continuous development and upgrades.

Headquarters structures. Generally unaffected (although leadership support and inputs are needed to guide requirements and budgeting decisions).

Requirements. Requirements need to be flexible, allowing the delivery of initially less capable system configurations. Subsequent requirements also should be flexible, allowing upgrade timing to be determined by technical maturity.

Budgeting. Budgeting processes are unaffected, but budgets need to be configured to reflect ongoing development and upgrades after IOC.

Downsides and trade-offs. Component-based upgrades can provide faster capabilities but might be constrained somewhat by the capabilities of the host system in which the upgrades are installed. This approach might also require opening up the interface if that is not already standardized or owned.

Design-Related

Descriptions

MOSA. This approach seeks to design systems with highly cohesive, loosely coupled, and severable modules that can be competitively acquired from independent vendors. This approach allows the DoD to acquire warfighting capabilities—including systems, subsystems, software components, and services—with more flexibility and with an ability to put component acquisition up for bidding again. MOSA involve a structure in which system interfaces share common, widely accepted standards with which conformance can be verified. The DoD is actively pursuing MOSA in the life-cycle activities of its MDAPs. In practice, MOSA systems engineering and integration can be difficult in that standards must be selected with potential associated compromises. MOSA might be faster initially if the standards selected facilitate plug-and-play of readily available components. This approach can also enable more-responsive and faster upgrades because the architecture and standards are known and components can be replaced to provide new capabilities.

Disaggregated architectures. This approach involves a system framework in which system components (and functionality) are geographically separated (for example, see Felt, 2013; Gruss, 2015; AFSPC, 2016; Risen, 2017; and Aerospace Corporation, 2018). In the realm of space, this refers to distributing computational and sensor capability across multiple craft. Disaggregated architectures might be more responsive in that new components can replace old ones or be added into the system by linking into the disaggregated interfaces. For example, a disaggregate satellite could add an improved image sensor by deorbiting the old sensor and flying a new sensor into its place.

Own the technical data (especially interfaces). This refers to the government taking care to retain IP and data rights of system interfaces, component designs, software code, and technology—or paying to acquire these rights when the benefits outweigh the costs. It is similar to the MOSA approach but might also involve not only an open architecture with nonproprietary interfaces but also owning the designs of parts of the system. This is intended to facilitate future competition and might enable government-led development, testing, and procurement of systems and add-ons. The technical data are owned by the DoD, so it has more flexibility in upgrading systems to respond to new needs or threats. If design rights are obtained, then production could be outsourced to alternative manufacturers to build to specifications to increase production rates or avoid any problems being experienced with the original manufacturer.

Conditions for Use

Table 2.11 lists the key conditions generally necessary for use of each approach and a short discussion of the reasoning behind the conditions.

Table 2.11. Key Conditions: Design-Related

Approach	Necessary Condition(s)	Reasoning
MOSA	• Suboptimal, modular architecture and standards are sufficient	Open-system architectures are applicable when the system performance does not require tightly optimized architectures that are difficult to modularize with standardized interfaces.[a] Many systems today are inherently modular with clearly defined interfaces as a result of their size and the number of developers involved, but modularity sometimes comes at the cost of compromising optimality or performance. Also, the MOSA approach attempts to use standardized interfaces, in which case the standard interfaces need to be sufficient (without modification) to provide the needed capabilities.

Approach	Necessary Condition(s)	Reasoning
Disaggregated architectures	• Suboptimal, modular architecture and standards are sufficient	Similar to the MOSA approach, a disaggregated architecture and system is applicable when the trade-off of benefits from disaggregation outweigh the reductions in system optimization and when the interface standards are sufficient to allow such disaggregation (for example, see Felt, 2013; Gruss, 2015; AFSPC, 2016; Risen, 2017; and Aerospace Corporation, 2018).
Own the technical data (especially interfaces)	• System modification, organic support, or alternative production contemplated	Paying to own the technical design data for systems should be considered when those data would be useful in the future, especially when future system modifications or alternative production sourcing are real possibilities. Otherwise, it might not be cost-effective to pay for the data designs. The first level of data to be considered is probably for the key system interfaces (i.e., to have an open-system architecture). The design of component elements (say, for alternative production of the components) are deeper considerations.

[a] See, for example, discussions of loosely and tightly coupled systems in Kossiakoff et al. (2011) and MITRE Corp. (2014).

Implementation Considerations

Modular Open-System Architectures (Allow Plug-and-Play Upgrades)

(Also see the more detailed discussion in Appendix A).

Acquisition strategies. This strategy relies on open standards to allow more-flexible trading of components from multiple vendors or to enable capability improvements more easily as technologies become available. The strategy also can increase competition (and reduce cost).

Acquisition processes. Processes focus on incremental development and fielding. Oversight processes are generally unaffected.

Program structures. Programs need sufficient systems engineering to design the architecture and select interface standards.

Headquarters structures. The MOSA approach requires broad, enterprise-wide coordination and might require suboptimal program-level decisions for portfolio- or enterprise-level success. This might imply the need for an enterprise-level governance structure.

Requirements. Requirements might need to reflect reduced or less efficient capabilities from an open, nonoptimized architecture to support an optimal enterprise. If incrementally focused, the requirements process in general would need to be flexible and feature thresholds or goals that can change over time to allow for expedient technology insertions.

Budgeting. The budget should reflect any needs to respond to emerging threats and technology developments through incremental upgrades. Note that future upgrades might come at a lower cost. Increased competition should reduce life-cycle costs in the long run.

Downsides and trade-offs. In addition to the implementation considerations, MOSA can result in less optimized systems because designs are constrained by the interface standards (e.g., module interactions are constrained by those interfaces). MOSA might also cost more in the short run because the prime contractor will be less motivated to incur immediate costs in anticipation of owning the long-term system "franchise" now that the DoD will have more flexibility to compete and control costs in the long term.

Disaggregated Architectures

Acquisition strategies. This strategy attempts to instill system flexibility by disaggregating components and facilitating plug-and-play both initially and in upgrades.

Acquisition processes. Oversight processes are generally unaffected.

Program structures. Programs need sufficient systems engineering to design the disaggregated architecture and select interface standards.

Headquarters structures. Unaffected.

Requirements. Requirements might need to reflect reduced or less efficient capabilities from an open, nonoptimized architecture.

Budgeting. Unaffected, although future upgrades might come at a lower cost.

Downsides and trade-offs. As discussed earlier, disaggregated architectures can result in less optimized systems because designs are constrained by the interface and its nonphysical connections.

Own the Technical Data (Especially Interfaces)

Acquisition strategies. This strategy needs to specify what technical data and IP to buy (or not buy) and whether the data should contain certain component designs and software in addition to the architectural interfaces. The strategy should also explicitly resolve other ownership factors, such as who owns the data generated by onboard diagnostic sensors.

Acquisition processes. Processes need to provide input and support for data-purchasing decisions.

Program structures. Programs need sufficient capabilities to exploit the acquired technical data.

Headquarters structures. Unaffected.

Requirements. Unaffected, although longer-range insights into future needs could help inform decisions on what technical data to purchase.

Budgeting. Purchasing of technical data can shift costs earlier in a program. Buying technical data for future competitions is often expensive and is usually affordable only if it is part of the original source selection.

Downsides and trade-offs. This approach might cost more in the short run because the DoD will have to pay for needed IP and the prime contractor will be less motivated to incur immediate costs in anticipation of owning the long-term system "franchise" now that the DoD will have more flexibility to compete and control costs in the long term.

Testing-Related

Descriptions

Test to acceptance. This approach involves testing by users to determine whether the system meets their operational requirements sufficiently to warrant acceptance and production. It is sometimes described as "good enough" testing. This approach can be faster because the user might accept the system before all normal testing is completed. It might increase risks in that all normal testing might not be completed, but the user is making an explicit trade-off of speed for risks using the information from tests that have been completed.

Operational testing in actual operations. Testing in operations (instead of a simulated environment) can speed initial capabilities to the warfighter in parallel with continued development and refinement. Usually such testing is limited to cases in which safety and operational risk are negligible—often when systems are in operational testing beyond development (although "Agile" development might include developmental testing in operations). Both traditional JCIDS operational test and evaluation phases and the (J)UONs and (J)EONs have formal operational testing of systems to meet specific, contractually mandated requirements. Examples are the joint surveillance and target attack radar system (JSTARS) and the Predator drone.

Conditions for Use

Table 2.12 lists the key conditions generally necessary for use of each approach and a short discussion of the reasoning behind the conditions.

Table 2.12. Key Conditions: Testing-Related

Approach	Necessary Condition(s) [a]	Reasoning
Test to acceptance	• Flexible acceptance levels for requirements	Testing to user acceptance (as opposed to a fixed set of requirements or to tests set by the operational test community) can be applicable when requirements (and thus acceptance by the user) are flexible and tradable. In other words, the user determines sufficiency and acceptability during development and testing rather than through a priori testing thresholds established at the beginning of the program. Test to acceptance is often applied for (J)UONs and (J)EONs (e.g., personal communication, JRAC).
Operational testing in actual operations	• Operators can tolerate unknown risks of system failure (that are not safety-related or mission critical) • Technology relatively mature	Conducting operational testing in actual operations is more common when the operators are willing to accept uncertain system performance (e.g., when the failure of an untested system would not cause significant safety concerns or put the mission success in jeopardy).

[a] Generally, each approach needs all the features listed (not just one).

Implementation Considerations

Test to Acceptance

Acquisition strategies. Testing to user acceptance (rather than to specific performance levels set early in the program) would need to be explicitly approved in this strategy.

Acquisition processes. Testing to acceptance affects how developmental and operational testing processes are conducted, including ensuring that users are consulted.

Program structures. Unaffected, except that user outreach mechanisms must be established.

Headquarters structures. Unaffected, except that the approach must be approved by either policy or MDA.

Requirements. Requirements need to lay out broad objectives set out initially but need to be flexible to reflect user acceptance during testing.

Budgeting. Unaffected, although the degree of user acceptance introduces some uncertainty in how much funding might be needed in the end.

Downsides and trade-offs. As mentioned earlier, risks are increased as accepted by the user (hopefully explicitly but perhaps unknown or insufficiently considered by the user).

Operational Testing in Actual Operations

Acquisition strategies. Performing operational testing in an actual (versus simulated) operational environment is an explicit element in the strategy.

Acquisition processes. Testing processes need to incorporate any special operational considerations (e.g., ensuring that the system under testing does not introduce safety or operational failure concerns).

Program structures. Unaffected, except that programs must have the capacity to reach out and interface with the operational community.

Headquarters structures. Unaffected, except that headquarters and MDA must approve operational engagements, ensuring that safety and operational effects are minimal and acceptable.

Requirements. Unaffected.

Budgeting. Unaffected.

Downsides and trade-offs. Testing in operational environments is limited if safety, security, or mission-assurance concerns prevail or might impose some of these concerns if testing proceeds.

Rapid-Fielding

Descriptions

    Rapid fielding: MTA (Section 804). The other MTA capability is rapid fielding. This approach is intended to use readily available, proven technologies to field production quantities of new or upgraded systems with minimal development required. Its goals are to begin production within six months and complete fielding within five years of an approved requirement (10 USC §2302 Historical and Revision Notes). This concept relies more on COTS or GOTS than on development of new technologies (as does rapid prototyping). Also see the rapid fielding part of Section 806 (10 USC §2447a–e) in the earlier section on technology development–related approaches.

(J)UONs or (J)EONs. By policy, the "DoD's highest priority is to provide warfighters involved in conflict or preparing for imminent contingency operations with the capabilities urgently needed to overcome unforeseen threats, achieve mission success, and reduce risk of casualties" (DoDD 5000.71). The objective is to deliver these capabilities quickly, preferably within days or months (DoDI 5000.02, Enclosure 13) but no longer than two years from the validation of the urgent need (DoDD 5000.71). The process for fulfilling these needs is within the traditional rapid acquisition process governed by DoDD 5000.71, the Warfighter Senior Integration Group, and DoDI 5000.02, Enclosure 13. JUON and JEON requirements are validated by the Chairman of the Joint Chiefs of Staff under the JCIDS process, but MDA is usually within a Component. Component-specific UON and EON requirements, program execution, and MDA are within the Component.

Conditions for Use

Table 2.13 lists the key conditions generally necessary for use of each approach and a short discussion of the reasoning behind the conditions.

Table 2.13. Key Conditions: Rapid-Fielding

Approach	Necessary Condition(s)[a]	Reasoning
Rapid fielding: MTA (Section 804)	• Budgetary resources obtainable • Production can be quick • Technology relatively mature	Rapid-fielding authorities under MTA apply only when production can begin within six months, and fielding can be completed within five years, of the development of an approved requirement (Section 804; 10 USC §2302 Historical and Revision Notes). These time constraints imply that needed budgetary resources can be obtained quickly (i.e., there is probably not enough time to request new funding through the normal POM process), that production can start very soon, and thus that the technologies involved should be relatively mature.
(J)UONs or (J)EONs	• Budgetary resources obtainable • Production can be quick • Requirements might be tradable given cost, schedule, or other issues that emerge during acquisition • Requirements urgent or emerging (UONs or EONs) • Technology relatively mature	JCIDS-based process for acquiring urgently needed capabilities requires completion within two years.

[a] Generally, each approach needs all the features listed (not just one).

55

Rapid Fielding: MTA (Section 804)

In addition to the general comments on prototyping above, the following specifics apply to the Section 804 MTA rapid prototyping (Pub. L. 114-92, 2015; 10 USC §2302 Historical and Revision Notes; USD[A&S], 2018b; SAF/AQ, 2018). Below are highlights of what the Secretary of Defense might provide in the guidance. However, specifics are evolving, so we highly recommend consulting the latest guidance from OSD and the military department in question.

Acquisition strategies. MTA rapid fielding (Section 804) programs must "begin production within six months and complete fielding within five years of the development of an approved requirement" (Section 804(b)(2)).

Acquisition processes. See the earlier section on MTA rapid prototyping (Section 804).

Program structures. See the earlier section on MTA rapid prototyping (Section 804).

Headquarters structures. See the earlier section on MTA rapid prototyping (Section 804).

Requirements. Approval of requirements are exempt from the JCIDS process.

Budgeting. Section 804(d)(2) specifies that the "Secretary of each military department might establish a military department-specific fund (and, in the case of the Secretary of the Navy, including the Marine Corps) to provide funds, in addition to other funds that might be available to the military department concerned, for acquisition programs under the rapid fielding and prototyping pathways established pursuant to this section. Each military department-specific fund shall consist of amounts appropriated or credited to the fund."

Downsides and trade-offs. The use of MTA authorities comes with the specific cost and schedule limitations and management procedures discussed above. Also, the implementation policies and guidance for MTA are in flux, including what degree of oversight and data tracking are required (compared with, say, prototyping efforts outside MTA). In addition, if MTA rapid fielding is to extend beyond MTA timelines, then transitions to other acquisition approaches and programs might need to be developed and executed.

(J)UONs or (J)EONs

Acquisition strategies. Strategies are tailored to focus on rapid delivery of capabilities to the warfighter, including such elements as parallel processes, minimal development, testing to user acceptance, and emphasis on urgency. Generally, "if the desired capability cannot be delivered within 2 years, MDA will assess the suitability of partial or interim capabilities that can be fielded more rapidly. In those cases, the actions necessary to develop the desired solution might be initiated concurrent with the fielding of the interim solution" (DoDI 5000.02, Enclosure 13). This strategy generally complies with the items needed for ACAT II and III programs, but "a streamlined, highly tailored strategy consistent with the urgency of the need will be employed. Regulatory requirements will be tailored or waived. The tailored acquisition strategy should be relatively brief and contain only essential information" (DoDI 5000.02). The PM collaborates with the requesting operational commander or sponsoring user representative to determine whether operational prototypes are pursued (DoDI 5000.02). The acquired system might, or might not, be transitioned to a program of record.

Acquisition processes. The process for (J)UONs and (J)EONs is laid out in the DoDD 5000.71 and Enclosure 13 of DoDI 5000.02. It involves tailored and streamlined program processes and oversight.

Program structures. Unaffected.

Headquarters structures. Depending on the program streamlining, the PM might report directly to MDA and the requirements validation authority.

Requirements. The Chairman of the Joint Chiefs of Staff "[s]erves as the validation authority for JUONs and JEONs and establishes procedures to ensure the timely validation of needs" (DoDD 5000.71) in accordance with the JCIDS (CJCSI 5123.01H). Validation of Component-specific UONs and EONs is done by the Component.

Budgeting. DoDI 5000.02 specifies that "Generally, funds will have to be reprioritized and/or reprogrammed to expedite the acquisition process. If a capability can be fielded within an acceptable timeline through the normal Planning, Programming, Budgeting, and Execution System, it would not be considered appropriate for urgent capability acquisition. . . . Funding for the acquisition program might be in increments over the program's life cycle."

Downsides and trade-offs. As with MDA, if (J)UON or (J)EON capabilities are needed in the long term, then transitions to other acquisition approaches, programs, and sustainment might need to be developed and executed.

Contracting-Related

Here, we focus on some broader concepts and approaches instead of the full set of contracting-related tools available in and outside the FAR, such as micropurchasing and others in common use by the contracting community.

Descriptions

Indefinite-delivery contracts (e.g., IDIQs). These are sometimes referred to as task-order (for services) or delivery-order (for supplies) contracts (for services and supplies, respectively; see FAR 16.501-1, 2019). IDIQs are often used for on-call service contracts (such as engineering) and are highly flexible. IDIQs can be as FAR contracts or under Other Transaction Authorities (OTAs) (FAR 16.501-2, 2019; GAO, 2018b). IDIQs can be faster in that the contract vehicle is already in place when a specific acquisition need is identified, avoiding a full source selection and contract negotiations. Some level of competition might be instilled between contractors on an IDIQ contract, but those competitions are simpler than starting a new contract from scratch.

LTPSI. This approach involves keeping a contractor on a specific program as long as the basic satellite architecture (the *bus*) remains essentially the same, thus avoiding additional lengthy solicitations and source selections and possible loss of programmatic and institutional knowledge throughout this period. Changing contractors in a program might result in delays to conduct a source selection, unnecessary systems integration challenges, and learning on the part of the new contractor with associated delays. In practice, changes in the prime systems integrator on major programs rarely happen (Table B.2).

Noncompetitive contracting: J&As. J&As for less than full and open competition under FAR Part 6 (2019) might speed acquisition of a known capability to a sole source. Award authority and protections are based on contract size. See also the noncompetitive part of Section 806 (10 USC §2447a–e) discussed in the earlier section on technology development–related approaches.

GSA schedule. These contracts—also referred to as Multiple-Award Schedules and Federal Supply Schedules—are "long-term governmentwide contracts with commercial firms providing federal, state, and local government buyers access to . . . commercial supplies (products) and services at volume discount pricing" (GSA, 2019). Schedules are intended to improve acquisition speed and reduce cost for the government because pricing is already negotiated and the GSA schedule is a preexisting contract vehicle. Schedules are also intended to provide a more direct link from contractors to the government (GSA, 2019).

Marketplaces. E-commerce portals are broader approaches beyond GSA schedules and the GSA's online marketplace (GSA Advantage!). The concept is to allow the DoD to buy commercial items online in marketplaces (such as Amazon, eBay.com, BestBuy.com, Jet.com, WalMart.com, and Shopping.com) in which commercial entities compete through online offerings. The FY 2018 NDAA directed phase I implementation of online marketplaces for the federal government (including the DoD) through GSA (Pub. L. 115-91, §846, 2017; GSA, 2018). The new GSA e-commerce portal is intended to streamline COTS purchasing up to the simplified acquisition threshold—$250,000 as of 2017 (41 USC §1901 amendments: Procurement Through Commercial E-Commerce Portals)—although GSA has proposed to also raise the micro-purchase threshold to $25,000 for GSA-approved portals to facilitate smaller purchases (GSA, 2018). Conceptually, other marketplaces could be constructed that fit within existing authorities; e.g., the Defense Logistics Agency's Electronic Catalog (for medical supplies, see Defense Logistics Agency, undated).

Bids open to consortium members only. Some contracts, including OTs, allow only members of their consortium of contractors to submit proposals. This can act as an incentive for good behavior and contractor performance. At the same time, it allows the government to ensure that it starts with a stable of suppliers that are already trusted to some degree. Time can be saved because bidders are already vetted and the consortium has a preexisting contracting structure in place. Consortiums might also contractually prohibit protests to the GAO.

Sole-source production following successful prototypes. Following the execution of a successful project, companies might negotiate for follow-on work. The award for such work or production can be sole-sourced as long as the initial contract (including OTs) was entered into using proper competitive procedures. (For example, see the OTAs of Section 806 [10 USC §2447a–e] discussed in the section on technology development–related approaches).

Sole-source production can be faster because it avoids a competitive solicitation and source selection. If a J&A is required that involves a lengthy process, then time savings might be minimal.

Custom IP arrangement via an OT. Custom arrangements in contracts (and OTs) can be made with regard to IP. Otherwise, the DoD might run into situations in which it cannot effectively modify or upgrade systems quickly or easily because it needs to quickly leverage commercial IP and capabilities but normal FAR arrangements can either (1) introduce government rights that are not acceptable to commercial entities that have invested in developing those capabilities or (2) slow down or complicate unique IP arrangements with one or more companies.

Team with commercial partner via an OT. Using an OT to work with a commercial partner can provide for more-flexible (non-FAR) contracts, IP agreements, and reduced administrative overhead. However, rules of competition often still apply.

Contracting that can avoid protests to the GAO (e.g., OTs). One potential advantage of OTs is that such transactions for prototypes are not open to protest to the GAO except in cases of misapplication of OTA. Bid protests to the GAO automatically stay contract awards and can induce delays in starting programs, so avoidance of such stays could be useful. However, protests can still be filed with the Court of Federal Claims (DoD Instruction 5000.02, 2018; USD[A&S], 2018a). Also, although the GAO avoids ruling within an OT, it will rule on protests involving whether the agency in question properly met the statutory requirements for using an OT in the first place (GAO, 2018a; Furin, 2018).

Use technical integration contract modifications or ECPs. Contracts can be modified relatively simply using technical integration contract modifications or ECPs to provide additional capabilities when both the government and the contractor agree to it. An example of this was Raytheon engineers modifying software on the SM-6 missile while still in development, allowing it to target surface objects in addition to the original purpose of targeting aircraft and missiles (LaGrone, 2016). Because the original contract vehicle already exists, modifications can be faster than creating a new contract vehicle (i.e., avoiding a new solicitation, source selection, and negotiation).

Contract schedule incentives (e.g., fixed-price incentive contracts, cost-plus incentive firm contracts, bonus for schedule). Contracts can be written to reward specific behaviors, such as meeting schedule performance goals or stretched schedule objectives. Incentives can be related to how much profit or fee the contractor earns on the contract (FAR Subpart 16.4, 2019).

Proven contractors only (might limit innovation). As described above, OT consortiums can use objective capability measures (e.g., responsiveness or proven ability to meet or beat schedules) to choose which contractors the government allows to participate. This can exclude contractors with poor track records or those too small to produce the sorts of items that the government is interested in (e.g., very small companies might not be able to design and build hyperspectral camera system for a new series of satellites). However, this can limit innovation. Also, source selections for FAR contracts, grants, and cooperative agreements can be constructed to contain objective capability measures and to account for any indication of poor past performance. This approach might also assist with managing risks.

Conditions for Use

Table 2.14 lists the key conditions generally necessary for use of each approach and a short discussion of the reasoning behind the conditions.

Table 2.14. Key Conditions: Contracting-Related

Approach	Necessary Condition(s)[a]	Reasoning
Indefinite-delivery contracts (e.g., IDIQs)	• Contracting areas known for future • Contractor prequalification possible or needed	IDIQ contracts are used "when the exact times and/or exact quantities of future deliveries are not known at the time of contract award" (FAR 16.501-2, 2019; GAO, 2018b). Although the exact timing and quantities are not known, this still implies a general belief that certain types of future needs are known (otherwise one would not bother setting up the IDIQ in the first place). Also, the process of selecting one or more performers for an indefinite-delivery contract mean that those awardees are prequalified. The reason to establish an indefinite-delivery contract is that it establishes a faster way to make specific orders once the indefinite delivery is established.
LTPSI	• Learning curve high or Infrastructure costs high • Sole-source (FAR 6.302) applies (for subsequent awards)[b]	The retention of an incumbent prime system integrator is generally considered when the cost of changing primes is high or other J&A criteria apply. Those costs can involve high infrastructure costs that would be incurred by a replacement prime contractor and technical expertise costs.
Noncompetitive contracting: J&As	• Sole-source (FAR 6.302) applies[b]	Sole-sourcing is restricted to the seven circumstances specified in FAR 6.302.
GSA schedule	• Able to use existing capabilities • Commonly needed contractor good or service	GSA schedules are established only when there is an expectation of federal need. Otherwise, the effort would not be worthwhile.
Marketplaces	• Able to use existing capabilities • Well-defined good or service	Marketplaces generally involve well-defined (as opposed to custom) goods and services. Existing capabilities also tend to involve relatively mature technology that can be marketed (as opposed to items involving extensive technology maturation).
Bids open to consortium members only	• Contracting areas known for future • Contractor prequalification possible or needed • Needs bidding approach beyond FAR • Needs strategic partnership beyond FAR	Like IDIQs, members-only bidding pools involve areas in which scope can be determined in advance to some degree and in which there is a basis for prequalifying bidders to join the pool. The use of OTAs to construct a non-FAR pool, however, also indicates a need beyond what is allowable in the FAR; otherwise, a standard FAR approach would be used. This approach can be done with traditional contracting, though OTs are popular for this purpose when adjustments from FAR contracting are helpful and appropriate.

Approach	Necessary Condition(s)[a]	Reasoning
Sole-source production following successful prototypes	• Prototype successful • Technology relatively mature	OTAs can allow sole-source production or rapid fielding following successful (mature) prototypes, bypassing normal FAR 6.302 competition roles. This approach can be done with traditional contracting, though OTs are popular for this purpose when adjustments from FAR contracting are helpful and appropriate.
Custom IP arrangement via an OT	• Commercial capability or technology of interest • IP protections needed beyond the FAR • OTA available	OTAs can allow custom arrangements to deal with IP protection concerns (especially with commercial entities). This is a consideration by DARPA when using OTA (DARPA, 2018). This approach can be done with traditional contracting, though OTs are popular for this purpose when adjustments from FAR contracting are helpful and appropriate.
Team with commercial partner via an OT	• Commercial capability or technology of interest • IP protections needed beyond the FAR • Needs strategic partnership beyond FAR • OTA available	OTAs can allow the government to team with a commercial partner that brings expertise or technologies of interest and helps fund the venture but also shares in the IP.
Contracting that can avoid protests to the GAO (e.g., OTs)	• OTA available	Because the GAO does not have bid protest jurisdiction regarding OTs for prototype projects (USD[A&S], 2018b), the delays associated with automatic stays on the contract award are not a concern. Agency-level protests and protests filed at the U.S. Court of Federal Claims are possible, and the court can issue a temporary restraining order, but it can be difficult and costs more (for example, see Lasky, 2007). Also, although the GAO avoids ruling within an OT, it will rule on protests involving whether the agency in question did not comply with the requirements of the relevant OT statutes (GAO, 2018a; Furin, 2018).
Use technical integration contract modifications or ECPs	• Budgetary resources obtainable • Able to use existing capabilities	Simpler technical integration contract modifications or ECPs can be used for rapid acquisition when relatively small adaptations from existing capabilities meet the needs.
Contract schedule incentives	• None (universal)	Well-designed and appropriate contract incentives to motivate accelerated acquisition can always be applicable.
Proven contractors only	• Contractor prequalification possible or needed • Contractor team highly skilled	The focus on proven contractors relies on the ability to make such determinations and when such highly experienced contractors exist. Note that this approach might limit innovation because good ideas can arise from new, untested sources (i.e., innovation might be traded for lower risk in other dimensions).

[a] Generally, each approach needs all the features listed (not just one).

[b] FAR 6.302 can be applied when at least one of the following conditions is met: "(i) Only one responsible source and no other supplies or services will satisfy agency requirements (substantial duplication of cost or unacceptable delay); (ii) Unusual and compelling urgency; (iii) Industrial mobilization; engineering, developmental, or research capability; or expert services; (iv) International agreement; (v) Authorized or required by statute; (vi) National security; or (vii) Public interest" (FAR 6.302, 2019).

Indefinite-Delivery Contracts (e.g., IDIQs)

Acquisition strategies. The use of indefinite-delivery contracts (FAR Subpart 16.5, 2019) could be an explicit element of this strategy (e.g., to establish a pool of qualified contractors that can be more rapidly awarded work as needed through already existing contract vehicles).

Acquisition processes. Unaffected.

Program structures. Unaffected.

Headquarters structures. Unaffected.

Requirements. Unaffected.

Budgeting. Unaffected.

Downsides and trade-offs. Indefinite-delivery contracts are common, but they do require planning to decide which contracts are needed and what their structures should be.

Long-Term Prime Systems Integrator

Appendix B contains additional discussion of practical considerations and insights into the LTPSI approach.

Acquisition strategies. Retaining a prime system integrator over the long term (e.g., over the life of a satellite system bus) could be an explicit strategy, but reasons for sole-sourcing the contract in subsequent years would have to meet one of the seven regulatory requirements set out in FAR 6.302 (2019; for more discussion, see Appendix B).

Acquisition processes. The process could be tailored to avoid subsequent competitions.

Program structures. Unaffected, although certain staff functions would not need to be expanded in what would have been future competitions.

Headquarters structures. Unaffected. MDA approvals should be possible under existing structures.

Requirements. Unaffected.

Budgeting. Unaffected, although cost savings might ensue.

Downsides and trade-offs. As far as we could tell from our admittedly limited case studies, the LTPSI approach has not become an a priori strategy but might evolve as justifications allow. See Appendix B for additional discussion.

Noncompetitive Contracting: J&As

Acquisition strategies. Pursuing noncompetitive elements to address near-term needs would need to be explicitly addressed and justified in the strategy.

Acquisition processes. Unaffected. Processes exist for requesting J&As.

Program structures. Unaffected.

Headquarters structures. Unaffected.

Requirements. Unaffected.

Budgeting. Unaffected.

Downsides and trade-offs. Given the predominance of competition as a major acquisition strategy within the DoD leadership, the FAR, and Congress, obtaining J&As can be difficult to obtain from the required DoD leadership. Success might vary depending on each leader's philosophy and acceptance of the PM's acquisition strategy, justifications, and data for the situation at hand.

GSA Schedule

Acquisition strategies. Acquiring items from a GSA schedule could be an explicit consideration in the strategy, especially if the components are significant to the acquisition.

Acquisition processes. GSA schedule implies a COTS or GOTS buy process.

Program structures. Unaffected.

Headquarters structures. Unaffected.

Requirements. Well-defined requirements are needed.

Budgeting. Unaffected.

Downsides and trade-offs. Although readily available, GSA schedule prices might be higher than those obtainable in a competitive or negotiated contract. For example, labor rates on the GSA schedule are often by categories of labor, not the specific costs of the individuals to be used on a contract. Depending on which GSA band an individual belongs to, his or her costs might be lower (or higher). This might result in higher prices to the DoD or time consumed in trying to negotiate a discount from the GSA schedule rates.

Marketplaces

Acquisition strategies. COTS components can be acquired under the simplified acquisition threshold using a commercial or government e-commerce portal. Although relatively small components can be acquired in this way, larger elements still fall within the limitations of 41 USC §1708, 15 USC §637, and FAR Part 5 (2019; also see Advisory Panel on Streamlining and Codifying Acquisition Regulations, 2018, p. 108).

Acquisition processes. Generally unaffected (although micropurchasing processes for contracting officers are affected).

Program structures. Unaffected.

Headquarters structures. Unaffected, unless Components decide to establish their own e-commerce marketplace across programs or acquired services.

Requirements. Depending on what is being acquired, requirements might need to be more flexible to accommodate system capabilities resulting from leveraging what is available through a COTS marketplace.

Budgeting. Unaffected, although the size limitation on GSA's forthcoming e-commerce portal under the FY 2018 NDAA (Pub. L. 115-91 §846) should be noted.

Downsides and trade-offs. Marketplaces can be efficient ways to quickly acquire COTS products while providing a level of competition. Still, they generally apply only to COTS components with the associated downsides discussed above for COTS components.

Bids Open to Consortium Members Only

Acquisition strategies. The use of a bidding consortium is an explicit strategy. What is acquired through the marketplace (e.g., system components or the entire system) and how this relates to system requirements and design will need to be decided.

Acquisition processes. If this is an organizational OT, then processes and elements will need to manage the consortium.

Program structures. If this is a program-specific OT, then program elements will need to manage the consortium.

Headquarters structures. If this is an organizational OT, then approval and oversight mechanisms are needed.

Requirements. Unaffected.

Budgeting. Unaffected.

Downsides and trade-offs. Restricting bids to preexisting members might mean that innovative approaches from nonmembers are missed. Also, if the bidding pool is very large, then the benefits and efficiency of this approach will depend on how the OT specifies the internal bidding process.

Sole-Source Production Following Successful Prototypes

Acquisition strategies. Direct, sole-sourced production is a strategy requiring either existing authorities (e.g., Section 806; see 10 USC §2447d(a)) or a J&A.

Acquisition processes. The program will need approval for either the use of Section 806 authorities or the J&A. The competition process for production would be bypassed.

Program structures. Unaffected, except that work to execute a competitive bidding would be replaced by work to obtain approvals for the sole-source production.

Headquarters structures. Unaffected, although specific approvals might be needed.

Requirements. Unaffected.

Budgeting. If Section 806 authorizations are used, up to $50 million in unused procurement funds from other accounts might be used to begin low-rate initial production of the rapid fielding project, although it is not yet clear whether congressional appropriators approve of this repurposing authority (10 USC §2447d(b)).

Downsides and trade-offs. Going directly to production of a specified prototype might mean that other potential opportunities in the marketplace are missed. Also, because competition is not in play, other mechanisms must be employed to ensure that a fair price is paid.

Custom IP Arrangement Via an OT

Acquisition strategies. OT authorities can allow for customization of IP rights if needed (e.g., when engaging commercial providers in a cost-sharing or consortium).

Acquisition processes. Unaffected, except that OTs are used.

Program structures. Unaffected.

Headquarters structures. Unaffected, although the OT specifics will need review and approval.

Requirements. Unaffected.

Budgeting. Unaffected.

Downsides and trade-offs. As with any OT, the risks of bypassing the lessons embedded in the FAR imply that greater care and planning are needed to ensure that the DoD and taxpayers get a good deal. In this case, the OT should ensure that the DoD gets reasonable rights commensurate with DoD investments and long-term needs.

Team with Commercial Partner Via an OT

Acquisition strategies. OTs can be used to establish custom teaming relationships with commercial partners beyond what is allowed in FAR arrangements.

Acquisition processes. Unaffected, except that OTs are used.

Program structures. The teaming relationship might affect what capabilities are needed in the program and how the program functions in relation with the commercial partner.

Headquarters structures. Unaffected, although oversight of the partnership will be needed.

Requirements. Partnerships might involve some compromises or expansions on what is developed.

Budgeting. Unaffected, although the partnership relationship might have budgeting effects.

Downsides and trade-offs. Again, as with any OT, the risks of bypassing the lessons embedded in the FAR imply that greater care and planning are needed to ensure that the DoD and taxpayers get a good deal. Details will depend on the kind of teaming arrangement envisioned.

Contracting That Can Avoid Protests to the GAO (e.g., OTs)

Acquisition strategies. At least some types of OTs (e.g., those for prototypes) can be used to avoid bid protests to the GAO, although protests can still be filed in federal court (USD[A&S], 2018b). Other strategies to avoid potential delays from bid protests could be employed (e.g., spending more time up front to review the source selection approach).

Acquisition processes. Unaffected, except that OTs are used.

Program structures. Unaffected.

Headquarters structures. Unaffected.

Requirements. Unaffected.

Budgeting. Unaffected.

Downsides and trade-offs. Avoiding bid protests in this fashion is limited. Also, there is a beneficial incentive from the bid protest process generally that motivates the DoD officers involved to make competitions fair and objective.

Use Technical Integration Contract Modifications or ECPs

Acquisition strategies. The use of contract modifications to add to or modify existing capabilities is time-honored but can limit the end result.

Acquisition processes. Unaffected, just uses a different process.

Program structures. Unaffected, although the program is configured around modifications rather than acquiring a whole new system.

Headquarters structures. Unaffected.

Requirements. The end capability might be more limited than for a new acquisition.

Budgeting. Unaffected, although needed resources might be lower.

Downsides and trade-offs. Modifications constrained the resulting capabilities (compared with the possibilities inherent in designing and producing a new system). Modifications also might incur any known cost, reliability, and performance issues of the existing system.

Contract Schedule Incentives

Acquisition strategies. The use of contract incentives (e.g., bonuses or higher profits or fees) for faster delivery is common, although perhaps less known than their use for controlling cost growth (Defense Procurement and Acquisition Policy [DPAP], 2016).

Acquisition processes. Unaffected, standard practices (see DPAP, 2016).

Program structures. Unaffected.

Headquarters structures. Unaffected.

Requirements. Unaffected.

Budgeting. Unaffected.

Downsides and trade-offs. Incentives are common practice. They might result in added costs to the DoD in higher profits or fees for better schedule performance, so the proposition codified in the incentive structures must reflect the cost–benefit values to the DoD.

Proven Contractors Only

Acquisition strategies. One strategy for mitigating risk is to focus on using only contractors that have a proven track record through objective source-selection criteria. Of course, this might limit innovation and high-risk, high-payoff opportunities.

Acquisition processes. Unaffected, standard FAR or OT procedures.

Program structures. Unaffected.

Headquarters structures. Unaffected, although the implications of risk avoidance need to be approved.

Requirements. This approach might result in less innovative solutions.

Budgeting. Unaffected.

Downsides and trade-offs. Restricting contracts to proven performers might miss innovation from new, unproven contractors.

Oversight

Descriptions

Senior board of directors (direct approval or oversight to PM). Direct reporting, oversight, and approval of major program activities by top decisionmakers in all domains

66

(requirements, financial management, acquisition, technology, and intelligence) can streamline and speed programs by avoiding multiple layers of staff and leadership review and enabling quick decisions by those with authority. Board members are often at the level of secretary; deputy, under, or assistant secretary; chairman or vice chairman of the Joint Chiefs of Staff; or service or vice chief. This approach has been used successfully by such organizations as the JRAC and DAF RCO. The challenge is that these individuals have limited time for direct oversight, usually limiting application to urgent or high-priority acquisitions. Also, some level of staff support might still be needed to help inform these decisionmakers, but such involvement can be a slippery slope to the very added reviews that the board was established to minimize in the first place.

Delegated board of directors (direct approval or oversight to PM). Like the senior board concept, delegated boards can be created if authority is granted to the individuals on the board without subsequent reviews above them. This might be especially applicable to lower-risk programs or those that do not require broader insights across areas that the delegated directors have privy to. Instead of membership existing at the secretary or chief level, it could, for example, be at the Senior Executive Service (SES) and O-6 to O-8 (colonel or captain to two-star general officer) levels.

Reinforced acquisition chain of command. One increasingly common approach to acquisition is to reduce staff and intervening decisionmaker layers within an acquisition chain of command. Making the organizations flatter increases individual decisionmaking ability, reduces the distance between decisionmakers and the people executing the program, and cuts administrative overhead delays. Also, bringing such support functions as contracting in house might allow a PM or organization to build highly skilled and responsive support functions, reducing the separation between various teams supporting and executing the program. Somewhat related, integrated product teams (IPTs) can be employed to streamline the support functions to the chain of command by resolving multifunctional equities in a parallel, coordinated fashion.

Limiting the number of programs reporting to each PEO. PEOs are a small group of officials within the DoD acquisition process. They might have only a single program under their authority (e.g., Joint Strike Fighter), or they might have dozens (e.g., Navy PEO Integrated Warfare Systems). One approach to making acquisition more rapid is to put fewer programs under PEOs, enabling PEOs to allocate more attention to each program and thus be more responsive (i.e., shortening access time).

MDA delegated to CAE. A traditional way to streamline the acquisition process and chain of command is to delegate MDA to the CAE,[11] eliminating most OSD oversight (and any

---

[11] The DAU Glossary defines the CAE as the

> Secretaries of the military departments or heads of agencies with the power of re-
> delegation. . . . The CAEs are responsible for all acquisition functions within their Components.
> This includes both the SAEs for the military departments and acquisition executives in other DoD

associated program schedule time). Although delegation can streamline the oversight and approval process, it can lead to missed opportunities for joint programs or other DoD-wide objectives and can allow more-parochial pressures to influence program decisions.[12] Delegation had been reserved for smaller or lower-risk programs, but the FY 2016 NDAA (Pub. L. 114-92, §825, 2015) changed the default MDA for MDAPs in a military department to be the SAE of the military department that is managing the program (i.e., the CAE of those Components) except in certain circumstances (for the latest details, see 10 USC §2430(d), as amended since FY 2016). Thus, although this delegation is part of the policy structure, we included it explicitly for longitudinal completeness because the pendulum on MDA delegation might swing the other way in the future.

MDA delegated to PEO. Further delegation of MDA from the CAE to the PEO shares some of the potential benefits and risks as delegation to the CAE. Recent efforts by some Components to flatten the chain of command and push decisionmaking authority to lower levels have involved delegation of MDA to the PEO (for example, see Air Force Instruction [AFI] 63-101/20-101, Table 1.1). Again, although this delegation is part of the policy structure, we included it explicitly for longitudinal completeness because the pendulum on MDA delegation might swing the other way in the future.

MDA delegated to PM. Still further delegation of MDA all the way down to appropriately qualified PMs shares some of the same potential benefits as delegation to the CAE and PEO but with increased risks. AFI 63-101/20-101 allows delegation of MDA for ACAT III programs to the PEO or "an appropriately qualified individual," such as a highly qualified PM (AFI 63-101/20-101).[13] This is an extension of the effort to push decisionmaking authority to the lowest level possible to flatten chains of command. Although this delegation is part of the policy structure, it might be the least stable because widespread PM expertise can be reduced by such measures as the implementation of Total System Performance Responsibility in the 1990s (for example, see Loudin, 2010).

Conditions for Use

Table 2.15 lists the key conditions generally necessary for use of each approach and a short discussion of the reasoning behind the conditions.

Components, such as the U.S. Special Operations Command (SOCOM) and Defense Logistics Agency (DLA), which also have acquisition management responsibilities. (Hagan, 2015)

The CAEs in the military departments are called SAEs and are the Assistant Secretary of the Army for Acquisition, Logistics, and Technology; the Assistant Secretary of the Navy for Research, Development and Acquisition; and the Assistant Secretary of the Air Force for Acquisition, Technology, and Logistics.

[12] For some examples of how parochialism can enter into decisions in various ways, see DAU, 2007; Nemfakos et al., 2010; Reed and Sereno, 2010; Rosenzweig, 2010; and Advisory Panel on Streamlining and Codifying Acquisition Regulations, 2018.

[13] ACAT III programs are smaller programs for which the eventual total expenditure for research, development, testing, and evaluation is estimated to be less than $185 million in FY 2014 constant dollars, or for procurement is less than $835 million in FY 2014 constant dollars (DoDI 5000.02).

Table 2.15. Key Conditions: Oversight

Approach	Necessary Condition(s)[a]	Reasoning
Senior board of directors (direct approval or oversight to PM)	• Requirement priority must be high	The use of direct senior approval and oversight boards (boards of directors) are usually limited to high-priority programs because they consume limited availability of senior leadership (e.g., at the level of assistant secretary and above).
Delegated board of directors (direct approval or oversight to PM)	• Risks low (generally)	A board of directors with lower-level membership that directly oversees a PM's program could be considered for lower-risk projects. Instead of secretary-level leaders, the membership could be delegated down to the PEO (for acquisition), a lower-level SES from the Comptroller's organization, and an O-6 or O-7 from the requirements community.
Reinforced acquisition chain of command	• None (universal)	Reinforced acquisition chains of command are generally applicable. If the chain of command directs a reduced staff involvement, then coordination and staff insight for the chain of command might be reduced, but there is no obvious condition under which this might be acceptable or appropriate.
Limiting the number of programs reporting to each PEO	• None (universal)	Limiting the number of programs that report to the PEO is universally applicable. This approach can allow for faster attention for approvals and on issues. It might be especially applicable when risks are higher and need more attention, but that condition is not required.
MDA delegated to CAE (the default)	• Joint issues largely absent	If programs have few or no joint issues, then delegation to the Component might be a reasonable option because the need for OSD to intervene and resolve them is reduced. Programs can still be joint, but our point here focuses on issues, not just jointness. Note that the SAE is MDA by default under law as of 2018.
MDA delegated to PEO	• Risks low (generally) • Joint issues largely absent	Delegation of MDA to the PEO can be considered when program risks are generally low and when joint issues are not significant.
MDA delegated to PM	• PM must be exceptionally experienced and skilled • Small dollar–value program • Risks low (generally) • Joint issues largely absent	Delegation of MDA to the PM can be considered when program risks are generally low, joint issues are not significant and the PM is highly skilled (and thus able to ensure that acquisition strategies are creative, that department goals are being achieved, and that problems are resolved).

[a] Generally, each approach needs all the features listed (not just one).

Implementation Considerations

Senior Board of Directors (Direct Approval or Oversight to PM)

Acquisition strategies. Application is usually limited to urgent or high-priority acquisitions. Thus, the strategy should reflect how the acquisition will accommodate rapid preparation and subsequent response to quick decisions by the board (e.g., streamlined acquisition and review processes, short schedules, rapid contracting approaches or available vehicles, quick acquisitions).

Acquisition processes. Reviews and approvals are at the level of secretary or of deputy, under, or assistant secretary with the acquisition chain of command on the board. Staff reviews should be minimal or eliminated.

Program structures. Unaffected, although capacity to support multiple levels of review should be significantly reduced or eliminated because the PM reports directly to senior leadership on the board.

Headquarters structures. This approach directly affects headquarters functions, bypassing separate lower-level reviews and requiring an ability for the PM to directly report to the entire acquisition chain of command simultaneously. Other senior leadership participation (e.g., from the service chief and Comptroller) facilitates faster resolution of any requirements or budgetary issues.

Requirements. Reviews and approvals of any requirements changes are handled directly at the level of the chairman or vice chairman of the Joint Chiefs of Staff or at the level of military service or vice chiefs, if part of the board. Staff reviews are minimal or eliminated.

Budgeting. Reviews and approvals of any budgetary changes are handled directly by the Comptroller; deputy; or, possibly, the department's general officer in charge of budget planning. Staff reviews are minimal or eliminated.

Downsides and trade-offs. As discussed, the biggest downside to this approach is the limited availability of senior leadership to participate on such boards.

Delegated Board of Directors (Direct Approval or Oversight to PM)

Acquisition strategies. Application is less limited than for senior boards, but the strategy should still reflect how the acquisition will accommodate rapid preparation and subsequent response to quick decisions by the board (e.g., streamlined acquisition and review processes, short schedules, rapid contracting approaches or available vehicles, quick acquisitions).

Acquisition processes. Reviews and approvals are handled directly at the level of SES and O-6 to O-8 (colonel or captain to two-star general officer), empowered by or delegated to officials in charge. For example, if authority still rests with MDA, then MDA could empower the PEO or MDA's deputy to chair the board and make prudent decisions for quick subsequent approval by MDA. Similar empowerment could happen for the requirements and Comptroller communities. As with the senior-board-of-directors approach, staff reviews should be minimal or eliminated.

Program structures. Unaffected, although capacity to support multiple levels of review should be significantly reduced or eliminated because the PM reports directly to senior leadership on the board.

Headquarters structures. This approach directly affects headquarters functions, bypassing separate lower-level reviews and eliminating reviews above the board. The headquarters structures must enable and facilitate direct reporting from the PM to the board (with any intervening parts of the acquisition chain of command having a presence on the board). Other senior leadership participation (e.g., from the service chief and Comptroller) facilitates faster resolution of any requirements or budgetary issues.

Requirements. Reviews and approvals of any requirements changes are handled directly by board members. Staff reviews are minimal or eliminated.

Budgeting. Reviews and approvals of any budgetary changes are handled directly by representatives from the Comptroller and general in charge of component budget planning. Staff reviews are minimal or eliminated.

Downsides and trade-offs. Although helping to mitigate the limited leadership time available in the senior board approach, this approach requires the availability of trusted delegates who can be empowered to make prudent decisions supported by leadership.

Reinforced Acquisition Chain of Command

Acquisition strategies. Depending on the propensities of superiors in the chain of command, more-innovative acquisition strategies might be possible given the usual tendency of staff in the bureaucracy to drive more-conservative approaches in their reviews.

Acquisition processes. This approach might directly affect staff involvement in the review and oversight processes, depending on directions from decisionmakers in the chain of command.

Program structures. Unaffected.

Headquarters structures. The headquarters reporting and oversight structures might be directly affected (i.e., staff involvement might be changed or reduced). Also, any intervening reviews by leadership not in the chain of command might be eliminated (e.g., reviews by executives or commanders that might be injecting themselves between the PM, PEO, or CAE).

Requirements. Depending on how it is implemented, this approach might also directly affect the requirements chain that the PM interfaces with (e.g., to update requirements as threats or technology change or to obtain requirements relief as issues arise).

Budgeting. Depending on how it is implemented, this approach might also directly affect the financial management chain that the PM interfaces with.

Downsides and trade-offs. If the chain of command directs a reduced staff involvement, then coordination and the staff insight provided to the PM, PEO, and CAE might be reduced or degraded.

Limiting the Number of Programs Reporting to Each PEO

Acquisition strategies. As with a reinforced acquisition chain of command, greater access to the PEO could enable a nimbler strategy and program approach and might allow for an improved ability to obtain leadership support for risk mitigation approaches.

Acquisition processes. Greater access to the PEO could reduce wait time.

Program structures. Unaffected, although this approach could improve PM mentoring and assistance as challenging issues arise.

Headquarters structures. This approach would probably lead to an increased number of PEOs and thus associated headquarters management of PEOs.

Requirements. This approach could increase the ability to get PEO support for requirements changes or to keep requirements stable.

Budgeting. This approach could increase the ability to get PEO support for new funding (if needed) or to keep budgets more stable.

Downsides and Trade-offs. For a fixed number of programs, this approach would mean that more PEOs are needed. Also, this approach might split related programs across multiple PEOs, making it harder for PEOs to resolve interdependencies.

MDA Delegated to CAE

Acquisition strategies. At the top level, ACAT is the primary driver of MDA, with some wiggle room at lower ACAT levels. Although not part of the strategy itself, MDA delegation affects how the PM thinks about program oversight and could enable a nimbler strategy and program approach and might allow for an improved ability to obtain leadership support for risk mitigation approaches.

Acquisition processes. Delegation directly affects the acquisition process, specifying who the final decision authority is and what levels of decisionmaker and staff review are involved. Reporting of general program status to OSD via the Component information systems would probably remain active.

Program structures. Unaffected, although capacity to support execution and milestone reviews to OSD are generally eliminated.

Headquarters structures. The headquarters reporting and oversight structures are directly affected as reporting structures and approvals by OSD are curtailed.

Requirements. Unaffected.

Budgeting. Unaffected.

Downsides and trade-offs. Shortened acquisition chains of command require exceptional abilities of subordinates and might mean that joint or cross-DoD equities might not receive sufficient attention.

MDA Delegated to PEO

Acquisition strategies. Although not part of the strategy itself, MDA delegation affects how the PM thinks about program oversight, could enable a nimbler strategy and program approach, and might allow for an improved ability to obtain leadership support for risk mitigation approaches.

Acquisition processes. Delegation directly affects the acquisition process, specifying who the final decisionmaking authority is and what levels of decisionmaker and staff review are involved. Reporting and monitoring of program status to the SAE—and general status reporting to OSD via the Component information systems—would probably remain active.

Program structures. Unaffected, although capacity to support execution and milestone reviews to OSD and the CAE—if present before—are generally eliminated.

Headquarters structures. The headquarters reporting and oversight structures are directly affected as reporting structures and approvals are curtailed above the PEO.

Requirements. Unaffected, although not having the CAE directly involved in program oversight might reduce CAE support when dealing with differences with the requirements community.

Budgeting. Unaffected, although not having the CAE directly involved in program oversight might reduce CAE support in budgetary prioritization battles.

Downsides and trade-offs. Shortened acquisition chains of command require that subordinates have exceptional abilities and might mean that joint or cross-DoD equities might not receive sufficient attention.

MDA Delegated to PM

Acquisition strategies. As with PEO delegation, although not part of the strategy itself, MDA delegation affects how the PM thinks about program oversight, could enable a nimbler strategy and program approach, and might allow for an improved ability to obtain leadership support for risk mitigation approaches. Risks might be higher because of the loss or significant reduction of PEO advice and mentoring, so added attention and risk mitigation might be warranted.

Acquisition processes. Delegation directly affects the acquisition process, eliminating much of the process involved in oversight decisions. Reporting and monitoring of program status to the PEO—and general status reporting to the CAE and OSD via the Component information systems—would probably remain active.

Program structures. Delegation to PMs should imply that only exceptionally qualified PMs should be assigned to the program. It might also imply that the program team should be exceptionally qualified to support PMs in their added decisional responsibilities. Also, capacity to support execution and milestone reviews to the PEO, CAE, and OSD (if present before) are generally eliminated.

Headquarters structures. The headquarters reporting and oversight structures are directly affected as reporting structures and approvals by the PEO, CAE, and OSD are curtailed.

Requirements. Unaffected, although not having the CAE and PEO directly involved in program oversight might reduce CAE support when dealing with differences with the requirements community.

Budgeting. Unaffected, although not having the CAE and PEO directly involved in program oversight might reduce CAE support in budgetary prioritization battles.

Downsides and trade-offs. Shortened acquisition chains of command require that subordinates have exceptional abilities and might mean that joint or cross-DoD equities might not receive sufficient attention.

3. Facilitating the Selection of Appropriate Approaches for Given Situations

One of the challenges in pursuing more-responsive acquisition approaches is simply figuring out where to start. As demonstrated by the length of Chapter 2, many approaches already exist, along with insights into their applicability. Selecting which approaches are appropriate given the situation and priorities at hand can be overwhelming. In this chapter, we describe how various factors can be used to select and finalize acceleration options.

Selection of Candidates Based on Necessary Conditions for Application

First, we considered the necessary conditions identified earlier for acceleration approaches. These conditions clarify what is needed before an approach can be pursued, and—because these conditions cannot always be met—they also serve to illustrate the limitations of many approaches. Box 3.1 summarizes the 49 necessary conditions identified across Chapter 2 for one or more acceleration approaches. Even this rather simplified list of conditions is complex, reflecting the diverse nature of the acceleration approaches and the attributes of acquisition that they seek to address.

Next, we mapped the necessary conditions against the list of acceleration approaches to examine the nature of the relationship. Figure 3.1 shows that this map is both large (essentially unreadable, in fact, for details in this scale) and sparse (i.e., even in this scale one can see that there are far more blank cells than cells with an X, which indicates that a condition applies to a specific approach). Thus, although one might consider using such a mapping to select appropriate approaches to consider (or, alternatively, to read through the lists and descriptions in Chapter 2), the task would be laborious (as the length of Chapter 2 might suggest).

There is not a clear framework for determining what degree of accelerated acquisition to use; rather, it is incumbent on senior, experienced personnel to determine what traditional acquisition controls are relevant and what might be unnecessary or tradable given other priorities.

In response, we programmed a simple spreadsheet to look up applicable approaches as the user selects which conditions apply to the acquisition situation at hand. Figure 3.2 provides a snapshot of the spreadsheet tool. As the user selects or unselects conditions on the left side, the spreadsheet tool looks up which approaches might be appropriate based on the map. In the example shown, the user selected seven existing conditions (budgetary resources obtainable, contractor PM exceptionally experienced and skilled, contractor team highly skilled, empowered decisionmakers, government PM exceptionally experienced and skilled, government staff highly skilled, and requirement priority high). Using the matrix mapping conditions to acceleration approaches, three options are identified (Skunk Works–like organizations, "crashing" the schedule, and senior board of directors for direct approval and oversight of the program). The conditions are necessary, but they are not sufficient for choosing an approach. Objective is a primary consideration, constrained by willingness to accept risk.

Box 3.1. Range of Key Conditions to Consider in Selecting More-Responsive Acquisition Approaches

Contracting
- Contractor prequalification possible or needed
- Needs bidding approach beyond FAR
- OTA available
- Sole-source applies[a]
- IP protections needed beyond the FAR

Financial
- Able to influence budgets
- Budgetary resources obtainable
- Budget needs relatively small in the near term
- Funding available in prototyping account

Process
- FAR exemptions consistently applicable
- Program small dollar value

Product
- Able to use existing capabilities
- System modification, organic support, or alternative production contemplated

Provider
- Government development
- Needs strategic partnership beyond FAR

Workforce
- Contractor team highly skilled
- Contractor PM exceptionally experienced and skilled
- Empowered decisionmakers
- Formal feedback loop for training or education
- Government staff capacity
- Government staff highly skilled
- Government PM exceptionally experienced and skilled

Requirements-related
- Commonly needed contractor good or service
- Contracting areas known for future
- Explore concept of operations
- Incremental capabilities useful
- Involve users in exploring capability options
- Joint issues largely absent
- Operators can tolerate unknown risks of system failure
- Requirements: flexible level of acceptance
- Requirements might be tradable given cost, schedule, or other issues that emerge during acquisition
- Suboptimal, modular architecture and standards are sufficient
- Threats are changing
- Well-defined good or service
- Requirement priority high
- Requirements urgent or emerging (e.g., [J]UONs, [J]EONs)
- Value determination needed
- Requirements approval authority at the component level

Risks (general)
- Security risks from foreign content are low
- Risks low (generally)

Technology-related
- Commercial capability or technology of interest
- Learning curve high or Infrastructure costs high[b]
- Production can be quick
- Prototyping possible in short time
- Prototype successful
- Risks of technology high [a]
- Software dominant
- Technical development needed and significant [a]
- Technology relatively mature

[a] FAR 6.302, 2019.
[b] These conditions would normally slow acquisition (or increase costs), but that fact might justify the use of an alternative approach or process that might then accelerate acquisition.

Figure 3.1. Illustrative Map of Necessary Conditions to Approaches

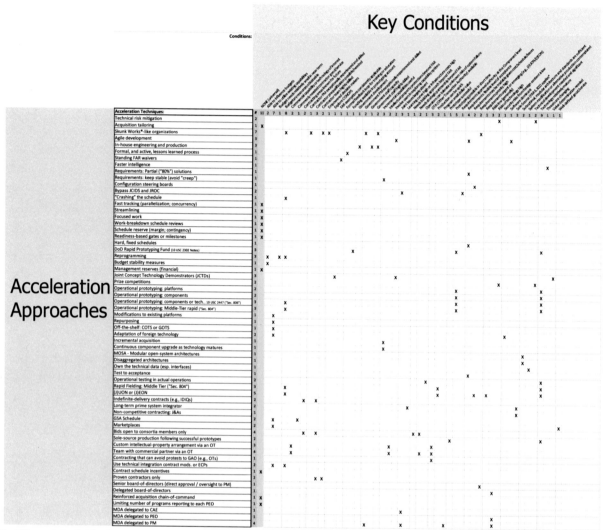

NOTE: Although individual rows and columns are essentially unreadable, the matrix in this figure illustrates the sparse nature of the detailed relationships between specific necessary conditions for using individual approaches. Cells with an X indicate that a condition applies to a specific acceleration approach. To read these details, see Chapter 2.

Figure 3.2. Snapshot of the Spreadsheet Lookup Tool for Approaches Based on Conditions

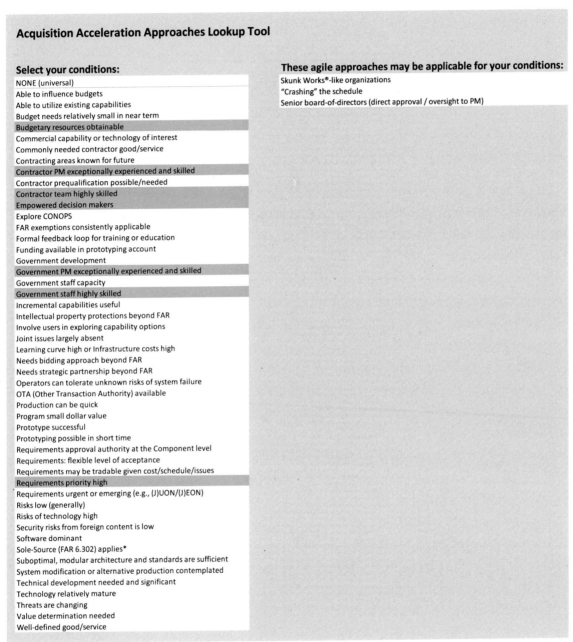

NOTE: This illustrates how the tool implements the relationship between the necessary conditions selected and the acceleration approaches discussed in Chapter 2. The user selects or deselects conditions on the left side, and the spreadsheet tool looks up which approaches might be appropriate and lists them on the right side.

Selection Based on Practical Implementation Considerations

Once candidate approaches are identified using the applicable conditions, the implementation considerations for each candidate should be reviewed to determine which approaches are viable. Again, those considerations feature attributes that relate to

- acquisition strategies

- acquisition processes
- program structures
- headquarters structures
- requirements
- budgeting
- downsides and trade-offs.

A brief discussion of these considerations for each approach is provided in Chapter 2.

Selection Based on Domains Involved

Acquiring a good or service involves more than an acquisition process; it depends on a validated requirement that details what to acquire and whether it is a critical need, and money also must be available. To accelerate development of physical systems, mature technology to build the system and intelligence to understand details of the threat are also needed. Thus, delays in these external domains can be as significant (or sometimes more significant) than delays that arise from within the acquisition system.

When selecting approaches, therefore, one should consider which of these domains the approaches can affect to mitigate as many impedances as possible.

Table 3.1 lists the domains that each approach addresses or affects. The details of involvement depend on a deeper discussion of the approach and the specific acquisition in question, but these tables are a starting point in finalizing selection of an approach after applying use conditions to build a list of candidate approaches. The domains are the key communities that affect acquisition: requirements, financial, technical, acquisition itself, and intelligence. Of note, intelligence affects requirements directly and the threat profile data used by the system during acquisition and sustainment.

For example, the general approach of pursuing systems that provide the most-effective capabilities in terms of cost and schedule while leaving the rest unfulfilled (the so-called 80-percent solutions) hinges on flexibility from the requirements community. Without this flexibility, the acquisition system is supposed to meet 100 percent of the requirements.

In another example, the relatively new VOLT reports are more-efficient ways to get current intelligence on the threat so that system designs and operating data can be updated accordingly. Although this helps ensure that relevant capabilities make it to the field, shifting requirements can add development time. However, such processes as VOLT can reduce the time penalty of ensuring that capabilities that reach the field are relevant.

The high-level Table 3.1 illustrates the breadth of domains involved in the approaches and shows that there are relatively few approaches that improve the interface between acquisition and intelligence systems. Recent efforts in the DoD's Better Buying Power initiative sought to improve this interface, but Table 3.1 demonstrates that more effort and research might be needed. Some unpublished DAF proposals have sought to improve the availability and currency of intelligence by creating organizational entities with the flexibility to adopt acquisition and contracting strategies to accept and act on these changes.[14]

Table 3.1. Domains Addressed or Affected by Approaches

Category	Approach	Requirements	Financial	Technical	Acquisition	Intelligence
High-level	Acquisition tailoring				X	
	Technical risk mitigation			X		
Combined	Skunk Works–like organizations	X	X	X	X	
	"Agile" development	X			X	
General	In-house engineering and production			X	X	
	Formal, active lessons learned process	X	X	X	X	X
	Standing FAR waivers	X	X		X	
Intelligence-related	Faster intelligence				X	X
Requirements-related	Requirements: Partial (80%) solutions acceptable to users	X				
	Requirements: Keep stable (avoid creep)	X				
	CSBs	X				
	Bypass JCIDS and JROC	X				
Schedule-based	Crashing the schedule		X	X	X	
	Fast-tracking (parallelization; concurrency)				X	
	Streamlining				X	
	Focused work				X	
	Work-breakdown schedule reviews				X	
	Schedule reserve (margin; contingency)				X	
	Readiness-based gates or milestones				X	
	Hard, fixed schedules	X			X	
Financial-related	DoD Rapid Prototyping Fund		X			
	Reprogramming		X			
	Budget stability measures		X			
	MRs (budget)		X		X	

---

[14] For example, such strategies as cost-plus contracts and MRs facilitate flexibility. Fixed-price contracts limit flexibility.

Category	Approach	Requirements	Financial	Technical	Acquisition	Intelligence
Technology development–related	Operational prototyping: new platforms			X	X	
	Operational prototyping: MTA rapid prototyping (Section 804)			X	X	
	Operational prototyping: Components					
	Operational prototyping: Components or technology (Section 806)			X	X	
	JCTDs	X		X	X	
	Prize competitions			X		
Reuse	Modifications to existing platforms	X				X
	Repurposing	X				X
	Off-the-shelf: COTS or GOTS	X			X	X
	Adaptation of foreign technology				X	X
Maturity-based modification	Incremental acquisition	X			X	X
	Continuous component upgrade as technology matures				X	X
Design-related	MOSA				X	
	Disaggregated architectures				X	
	Own the technical data (especially interfaces)				X	X
Testing-related	Test to acceptance	X				X
	Operational testing in actual operations	X				X
Rapid-fielding	Rapid fielding: MTA (Section 804)					X
	(J)UONs or (J)EONs	X				
Contracting-related	Indefinite-delivery contracts (e.g., IDIQs)					X
	LTPSI					X
	Noncompetitive contracting: J&As					X
	GSA schedule					X
	Marketplaces					X
	Bids open to members only					X
	Sole-source production following successful prototypes					X
	Custom IP arrangement via an OT				X	X
	Team with commercial partner via an OT				X	X
	Contracting that can avoid protests to the GAO (e.g., OTs)					X
	Use technical integration contract modifications or ECPs					X
	Contract schedule incentives					X
	Proven contractors only					X

Category	Approach	Requirements	Financial	Technical	Acquisition	Intelligence
Oversight	Senior board of directors (direct approval or oversight to PM)	X	X	X	X	
	Delegated board of directors (direct approval or oversight to PM)	X	X	X	X	
	Reinforced acquisition chain of command				X	
	Limiting the number of programs reporting to each PEO				X	
	MDA delegated to CAE (the default)				X	
	MDA delegated to PEO				X	
	MDA delegated to PM				X	

Selection Based on the Stage in an Acquisition's Life Cycle

Finally, the appropriateness of an approach might depend on the point at which an acquisition is within its life cycle. This consideration is hard to tie to normal acquisition phases (e.g., before a certain phase or milestone). For example, approaches that work best at the start of an acquisition effort might also be useful at the start of a modification, prototype, or experimentation involving an existing system in sustainment. Nevertheless, the following observations give insight to when an approach might be appropriate.

Some approaches involve the fundamental structure of the acquisition approach and the larger acquisition processes within which it operates. Thus, these approaches are essentially limited to the start of an acquisition:

- Skunk Works–like organizations
- in-house engineering and production
- "Agile" development
- MOSA
- disaggregated architectures
- (J)UONs or (J)EONs.

Note, however, that this constraint might not apply to modifications and upgrades to an existing system that could employ most of these approaches (except MOSA and disaggregated architectures, which involve the fundamental design of the system).

Other approaches involve technology development for a system (either during initial design or for a modification). Thus, if development has largely been completed, then these approaches might not be relevant:

- technical risk mitigation
- "Agile" development
- JCTDs
- operational prototyping: new platforms

81

- operational prototyping: MTA rapid prototyping (Section 804)
- operational prototyping: components
- operational prototyping: components or technology (Section 806)
- prize competitions.

The remaining approaches, although perhaps most effective at the start of an acquisition process, could conceptually be added to an ongoing acquisition to seek some benefits. For example, a CSB could be added to assist and guide a program in managing its requirements in the face of changing threats or external pressures. Also, MDA delegation during a program often occurs because risks are resolved.

4. Organizational Models: Illustrating How Acceleration Approaches Can Be Combined and Implemented

Organizations that pursue accelerated acquisition often use multiple approaches to accelerate their acquisitions. It is illuminating, therefore, to examine several DoD organizations to see which approaches they use in combination to address specific needs related to their mission matters. Thus, this chapter discusses the following agile organizations from our research:

- Big Safari
- U.S. Special Operations Command (USSOCOM)
- NRO
- DAF RCO
- OSD's JRAC
- Air Force Rapid Development Integration Facility
- DIU
- Army C5
- Lockheed Martin's Skunk Works (and similar concepts)
- Strategic Capabilities Office (SCO).

After discussing each organizational model, we list the individual acquisition approaches identified in the model. These organizational strategies for accelerated acquisition draw on multiple approaches and techniques. These lists are not necessarily exhaustive of all approaches used by these organizations, but they illustrate at least many of the approaches we identified and how different sets of approaches can be combined to assist the mission of different rapid acquisition organizations.

We then briefly examine how the combinations of approaches might affect six implementation areas (acquisition strategies, acquisition processes, program structures, headquarters structures, requirements, and budgeting) as well as any downsides and trade-offs. These practical considerations factor in which of the six areas must apply the acceleration approaches involved and deal with the necessary conditions for using those approaches, along with any additional elements that should be considered beyond those identified in the conditions lists. A table at the end of each organization description provides an overview of practical considerations for these models across the six dimensions, as well as any downsides and trade-offs. These considerations also help explain how these organizations combine acceleration approaches in their operational model. Finally, we highlight common themes and observations from these organizational case studies at the end of the chapter.

Big Safari

Big Safari began in 1952 as a program to procure specially modified aircraft to conduct surveillance and reconnaissance missions in Europe. Well-known examples of more-recent unclassified Big Safari programs are RIVET JOINT, COBRA BALL, COMPASS CALL, and COMBAT SENT. Big Safari is headquartered at Wright-Patterson AFB with associated facilities at Hanscom AFB and Greenville, Texas. Featured aircraft are the EC-130, RC-135, and unmanned aerial vehicles (Grimes, 2014).

Big Safari focuses on nondevelopmental upgrades to existing equipment. Most upgrades are COTS or GOTS, and upgrades are often made on a continuing basis to meet evolving threats and capability needs. Programs are typically driven by UONs and JUONs, and they are smaller in size then MDAPs. This flow of upgrades leads to mixed configuration fleets of aircraft. Despite this, Big Safari provides full life-cycle support.

The programs are typically ACAT II, ACAT III, or special-access programs. Big Safari operates under the FAR and JCIDS. When possible, it tailors waivers and processes. In particular, reporting and decision waivers are used to shorten the contracting process by several weeks. Most contracts are firm-fixed-price or cost-plus-fixed-fee. They do not have difficulty handling undefinitized contract actions,[15] which allows for greater flexibility and speed under FAR 16.603 (2019).

What sets Big Safari apart is primarily culture. Long-standing partnerships with industry have been developed, yet Big Safari can stop working with any company that becomes too difficult to work with, is unresponsive, or cannot embrace the culture of rapid acquisition. Contractors work hand-in-hand with Big Safari onsite. Users are brought in to provide rapid insight into operational needs and limitations. Unlike some other programs, this culture is not codified into a set of rules.

Risk-taking to accelerate acquisition is rewarded, and failures (when acting aggressively to pursue capability) are penalized to a much lesser degree than elsewhere—both for contractors and for the government PMs. Big Safari does not operate under a board of directors but reports to the PEO of intelligence, surveillance, and reconnaissance (ISR) and to MDA—which is SAF/AQ. The PEO of ISR accommodates the urgency of the Big Safari mission by limiting or reducing the amount of reporting required. At the congressional level, Big Safari works with staffers to reduce resistance to its model while ensuring that the urgency of specific acquisitions is understood. Rarely has Congress intervened in Big Safari programs, and oversight is similar to that for ACAT II and III programs.

Table 4.1 lists the individual acquisition approaches that are integrated in Big Safari's overall acquisition formula. Table 4.2 then briefly outlines how the combinations of these approaches

---

[15] An *undefinitized contract action* is "any contract action for which the contract terms, specifications, or price are not agreed on before performance is begun under the action" (DFARS, 2019, Subpart 217.7401).

might affect six implementation areas (acquisition strategies, acquisition processes, program structures, headquarters structures, requirements, and budgeting).

Table 4.1. Big Safari Acquisition Formula

Category	Approaches Used	Requirements	Financial	Technical	Acquisition	Intelligence
High level	• Acquisition tailoring				X	
Combined	• Skunk Works–like organizations	X	X	X	X	
General	• In-house engineering and production			X	X	
	• Formal, active lessons learned process	X	X	X	X	X
Intelligence	• Faster intelligence				X	X
Requirements	• Requirements: Partial (80%) solutions acceptable to users	X				
	• Bypass JCIDS and JROC (sometimes)	X				
Financial	• Budget stability measures		X			
Technology	• Operational prototyping: components					
Reuse	• Modifications to existing platforms	X			X	
	• Off-the-shelf: COTS or GOTS	X		X	X	
Mature mods.	• Continuous component upgrade as technology matures			X	X	
Testing	• Test to acceptance	X			X	
	• Operational testing in actual operations	X			X	
Rapid fielding	• Rapid fielding: MTA (Section 804)				X	
	• (J)UONs, (J)EONs	X				
Contracting	• Indefinite-delivery contracts (e.g., IDIQs)				X	
	• LTPSI				X	
	• Noncompetitive contracting: J&As				X	
	• Proven contractors only				X	

Table 4.2. Big Safari Implementation Considerations and Conditions

Acquisition Strategies	Primarily COTS or GOTS upgrades to existing aircraft
Acquisition Processes	Follows the FAR; tailored ACAT II and ACAT III processes; higher risk tolerance; work is onsite; ability to avoid contractors who fail to perform or integrate with in-house engineering teams
Program Structures	Programs specialized to facilitate rapid COTS or GOTS insertion onsite
Headquarters Structures	No effect
Requirements	(J)UON-driven; JCIDS; waivers encouraged and expedited
Budgeting	Work with Comptroller to quickly identify funding

U.S. Special Operations Command

The USSOCOM Commander has unique acquisition authorities and responsibilities. 10 USC §167 vests the USSOCOM Commander with the responsibility and authority for the development and acquisition of Special Operations–peculiar equipment and the authority to execute funds. The Commander has delegated those authorities to the USSOCOM Acquisition Executive, who leads SOF Acquisition, Technology, and Logistics (SOF AT&L). Congress also provided USSOCOM with the specific appropriation funding to support the development, acquisition, and sustainment activities for Special Operations–peculiar equipment. This funding helps the Command meet its unique, time-sensitive mission requirements.

SOF acquisition contracting directly supports SOF AT&L and other USSOCOM Joint Staff Directorates. SOF AT&L Contracting (SOF AT&L-K) awards command-wide, large-dollar special operations equipment and performance-based service contracts. It uses multiple contracting offices located throughout the continental United States and in overseas contingency environments. Each office provides support to USSOCOM PEO, Directorates, or Military Service SOF components or units.

USSOCOM can approve contract awards in days, if required. There are several reasons for this. First, it has delegated procurement authority and is generally procuring COTS or GOTS equipment. Second, USSOCOM's acquisition approval authorities are resident at the same headquarters location as the program and contracting teams. Lastly, SOF AT&L-K professionals participate in IPT meetings and other early planning meetings to help accelerate the approval process and minimize requirement changes before contract award. (The IPT concept is seen in several other programs described in this chapter.) This is the point at which requirements, based on user and operator inputs, are set. Because of the pace of operational needs, (J)UONs and (J)EONs are generally not used.

Collectively, these factors lessen the processing between the contracting officer and the approving official, resulting in expedited approvals (U.S. Special Operations Command, Special Operations Forces [Acquisition, Technology, and Logistics], undated).

Table 4.3 lists the individual more-responsive acquisition approaches that are integrated in USSOCOM's overall acquisition formula. Table 4.4 briefly outlines how the combinations of these approaches might affect six implementation areas (acquisition strategies, acquisition processes, program structures, headquarters structures, requirements, and budgeting).

Table 4.3. USSOCOM Acquisition Formula

Category	Approaches Used	Requirements	Financial	Technical	Acquisition	Intelligence
High level	• Acquisition tailoring				X	
Intelligence	• Faster intelligence				X	X
Requirements	• Requirements: Partial (80%) solutions acceptable to users	X				
Financial	• Budget stability measures		X			
Technology	• Operational prototyping: components					
Reuse	• Modifications to existing platforms	X			X	
	• Off-the-shelf: COTS or GOTS	X		X	X	
Testing	• Test to acceptance	X			X	
	• Operational testing in actual operations	X			X	
Rapid fielding	• (J)UONs, (J)EONs	X				
Contracting	• Noncompetitive contracting: J&As				X	

Table 4.4. USSOCOM Implementation Considerations and Conditions

Acquisition Strategies	Heavily COTS and GOTS reliant; acquisition decisions delegated to USSOCOM Commander level
Acquisition Processes	IPTs and leadership coordinate to produce requirements quickly and reduce COTS award time via internal approval authority
Program Structures	Multiple SOF AT&L-K offices procure and field equipment to operators quickly, each of which has the organic capability to perform the contracting function; acquisition decisions at lowest level possible, even for expensive equipment
Headquarters Structures	Acquisition Executive at headquarters level is unusual
Requirements	Internally driven by user demand; generally, requirements do not go through JCIDS, (J)UON, or (J)EON processes
Budgeting	ACAT II or III authority to USSOCOM Commander; awards are performance-based

National Reconnaissance Office

The NRO is a DoD member of the Intelligence Community responsible for providing satellite-based imagery intelligence (IMINT), signals intelligence (SIGINT), and measurement and signature intelligence (MASINT). Historically, these included the CORONA and KEYHOLE series satellites. The NRO designs, builds, and operates these satellites. The NRO has a substantial budget and leverages contractor support. The existence of the NRO was classified until 1992.

The NRO employs several strategies for developing capabilities rapidly, depending on how quickly the capability is needed. For extremely urgent, mission-focused needs (within a few days to a few months), the NRO relies on existing platforms in orbit to achieve the mission. When such a need arises, it uses available contract engineers to find new or novel ways to use a platform (or multiple platforms in conjunction) to obtain the intelligence needed. Generally, these are software-based solutions developed in house.

This short-term concept of adapting, improvising, and overcoming challenges has interesting considerations. Sometimes, it is not possible to create the capability needed with the assets already in orbit. This concept also relies on engineering staff that have worked with the orbital systems in question for years, possibly even decades. The NRO, therefore, effectively employs a kind of LTPSI approach, at least in the short run.

In the medium term, the NRO can fund innovative technologies at the director's discretion through the Director's Innovation Initiative (NRO, undated-b). Funding for these initiatives is supplied every year, providing some flexibility to meet emerging needs. Rarely are these initiatives driven by the formal JCIDS-based (J)UON and (J)EON process. The goal is to get these new technologies into space in two years or less (Werner, 2018).

The other NRO medium-term acquisition strategy employs COTS satellite architectures and small, commercially built launch vehicles to get specific capabilities (including new technologies developed via the Director's Innovation Initiative) into orbit as quickly as possible (Levinson, 2018). This strategy of small, fast, COTS capability is part of a philosophy that emerged in the wake of the Future Imagery Architecture program failure.[16] As a result, NRO often accepts moderate solutions that can be delivered quickly rather than waiting for complete solutions that take more than ten years, for example. The NRO also uses these small satellites as platforms to test new technologies that might be incorporated into larger systems in the future.[17]

Much of this philosophy is also applied to larger programs with a more traditional development cycle. The NRO considers commercial rockets (e.g., the Falcon V) to promote competition, availability, capability, and cost reduction. The development of satellites will attempt to avoid some of the failures of the 1990s and 2000s by accepting less capability in exchange for faster acquisition time frames (Berkowitz, 2015). Of the agencies we interviewed, the NRO was the most vocal about consciously considering (formally or informally) an LTPSI strategy to accelerate the development and fielding of capabilities—whether in the short or long term.

Table 4.5 lists the individual more-responsive acquisition approaches that are integrated in the NRO's overall acquisition formula. Table 4.6 briefly outlines how the combinations of these approaches might affect six implementation areas (acquisition strategies, acquisition processes, program structures, headquarters structures, requirements, and budgeting).

---

[16] In 1999, Boeing underbid incumbent Lockheed Martin to develop a new generation of radar imaging satellites, despite Boeing's limited experience in the field. The Future Imagery Architecture program was canceled in 2005, after $10 billion had been spent, with cost overruns between $4 billion and $5 billion, making it the most expensive failure of a satellite program in U.S. history. It is frequently regarded as a cautionary tale about switching contractors.

[17] Based on discussions with the NRO Acquisitions Office in February 2018.

Table 4.5. NRO's Acquisition Formula

Category	Approaches Used	Requirements	Financial	Technical	Acquisition	Intelligence
High level	• Acquisition tailoring				X	
	• Technical risk mitigation			X		
General	• Formal, and active, lessons learned process	X	X	X	X	X
Intelligence	• Faster intelligence				X	X
Requirements	• Requirements: Partial (80%) solutions acceptable to users	X				
	• Requirements: keep stable (avoid creep)	X				
Schedule	• Hard, fixed schedules	X		X		
Financial	• Reprogramming		X			
	• Budget stability measures		X			
	• MRs		X	X		
Technology	• Operational prototyping: platforms			X	X	
	• Operational prototyping: components					
Reuse	• Modifications to existing platforms	X		X		
	• Repurposing	X		X		
	• Off-the-shelf: COTS or GOTS	X		X	X	
Mature mods.	• Incremental acquisition	X		X	X	
Contracting	• LTPSI (if appropriate)				X	
	• Proven contractors only				X	

Table 4.6. NRO Implementation Considerations and Conditions

Acquisition Strategies	Strategies depend on what is needed and how soon (short to medium term: small, commercial buses; long term: partial near-term solutions followed by upgrades when technology matures); COTS launch vehicles
Acquisition Processes	Improvise, adapt, overcome mindset for immediate challenges
Program Structures	Director's Innovation Initiative to fund new technologies into space in less than 2 years; local control of software via contractors
Headquarters Structures	Relatively short chain-of-command
Requirements	Partial solutions acceptable; flexible contract requirements for engineer duties; rarely use (J)UONs
Budgeting	Annual budget for Director's Innovation Initiative to meet emerging needs and opportunities; small overall[a]

[a] The FY 2020 Broad Area Announcement for the NRO Director's Innovation Initiative says that it "is expected that multiple awards will result from proposals received, at a maximum funding level of $500,000.00 with a nine (9) month period of performance" (NRO, 2019).

Department of the Air Force Rapid Capabilities Office

The Secretary of the Air Force established the RCO on April 28, 2003, in response to the inflexibility and slow response timelines associated with the normal acquisition cycle. The office's guiding principles were later codified by a charter in 2010 and updated in 2018. Currently, the DAF RCO reports directly to a Board of Directors, including the USD(A&S), the Secretary of the Air Force, the Chief of Staff of the Air Force, the Chief of Space Operations, the Under Secretary for Research and Engineering (USD[R&E]), and SAF/AQ. The office responds to Combat Air Force and Combatant Command requirements (USAF, 2009) and adopts many of its philosophies from the Lockheed Skunk Works model (Tirpak, 2018).

The DAF RCO develops technologies on an accelerated timeline. It works across the government and leverages existing technologies to speed capabilities into service that meet the urgent needs of the warfighter. The office also conducts experiments into advanced processes, methods, and techniques while performing independent operational and technical assessments of integrated weapon systems and combat support systems.

The DAF RCO's chain of command is shorter than in most other DoD organizations, with the specific intent to reduce the amount of bureaucracy and red tape involved in program execution. Additionally, the DAF RCO has found success by pushing authority in the acquisition process to a lower level. The office is typified by small teams consisting of legal, engineering, and financial experts; users; PMs; and contracting officers. It relies heavily on integrated warfighter involvement as a driver for new capabilities and initiatives.

The DAF RCO has the flexibility to allocate certain sources of stable, year-over-year funding to pursue new initiatives and capabilities identified by the warfighter and approved by its Board of Directors. Additionally, requirements are kept stable to avoid drawn-out acquisition timelines. The DAF RCO follows the FAR and DFARS while adhering to the tailored philosophy of DoDI 5000.02.

Similar to the NRO, the DAF RCO leverages mature technology in COTS and GOTS to accelerate acquisition. It also focuses on rapid point solutions over consortium-driven products, which tend to wait for 100-percent solutions (Walden, 2016).

The DAF RCO recruits highly experienced, self-motivated, and creative individuals who seek innovative ways to move programs forward. For example, although contracting has been a source of frustration for many acquisition programs, interviews indicated that the DAF RCO contracts office seeks only the highest-qualified individuals who can leverage all existing authorities to find the fastest way to get work on contract. These interviews also indicated that it is hard to assemble such a high-quality team across the DAF.

Notable DAF RCO programs include the X-37B Orbital Test Vehicle, the Common Mission Control Center, and the B-21 Raider Long-Range Strike Bomber. Although the B-21 has yet to achieve IOC, the DAF RCO is credited with the success of the B-21 program (Erwin, 2018b). As a result, new organizations (such as a new Army RCO and the Space RCO) are trying to model their structures and processes after the DAF RCO's.

Table 4.7 lists the individual responsive acquisition approaches that are integrated in the DAF RCO's overall acquisition formula. Table 4.8 briefly outlines how the combination of these approaches might affect six implementation areas (acquisition strategy, acquisition process, program structure, headquarters structure, requirements, and budgeting).

Table 4.7. The DAF RCO's Acquisition Formula

Category	Approaches Used	Requirements	Financial	Technical	Acquisition	Intelligence
High level	• Acquisition tailoring				X	
Combined	• Skunk Works–like organizations	X	X	X	X	
Requirements	• Requirements: Partial (80%) solutions acceptable to users	X				
	• Requirements: keep stable (avoid creep)	X				
Schedule	• Hard, fixed schedules	X			X	
Financial	• Budget stability measures		X			
Technology	• Operational prototyping: platforms			X	X	
	• Operational prototyping: components					
Reuse	• Modifications to existing platforms	X			X	
	• Off-the-shelf: COTS or GOTS	X		X	X	
Testing	• Test to acceptance	X			X	
	• Operational testing in actual operations	X			X	
Rapid fielding	• (J)UON, (J)EON	X				
Contracting	• Noncompetitive contracting				X	
Oversight	• Senior Board of Directors (direct approval or oversight to PEO)	X	X	X	X	
	• Reinforced acquisition chain of command				X	

Table 4.8. DAF RCO Implementation Considerations and Conditions

Acquisition Strategies	COTS, GOTS if possible; rapid point solutions; partial solutions acceptable
Acquisition Processes	Follows the DoDI 5000.02 and associated directives and instructions; warfighter involvement at all phases; "A Team" of contracting officers
Program Structures	Shorter chains of command; small teams that include legal, engineering, and financial experts; contracting officers; PMs; and users
Headquarters Structures	Use of a Board of Directors bypasses standard low-level staff processes; able to bypass additional processes because of classification
Requirements	Response to Combatant Commander Urgent requirements; (J)UON, (J)EON process
Budgeting	Follows FAR and DFARS; has both the flexibility to pursue projects (with Board of Directors approval) and a relatively high degree of security in annual funding

OSD's Joint Rapid Acquisition Cell

According to the JROC Charter, the OSD JRAC collaborates with the Joint Staff Gatekeeper and Functional Capabilities Boards in the review of proposed JUONs and JEONs prior to validation. The OSD JRAC became responsible for developing and implementing policies for the Warfighter Senior Integration Group (SIG), as well as assigning working groups subordinate to the SIG, in order to speed capabilities to fight in Afghanistan and against the Islamic State (DoDD 5000.71). The JRAC director serves as the senior official within OSD responsible for making recommendations to the Secretary of Defense on the use of rapid acquisition authority (Pub. L. 107-314, §806(c–d)). The OSD JRAC reports to USD(A&S), who also acts as the chair of the Warfighter SIG (DoD Manual 5000.78).

The JRAC was established to meet JUONs and JEONs in less time than the standard defense acquisition process. The JRAC receives and evaluates validated JUONs and JEONs, collaborating with the Joint Staff J-8 and the respective Combatant Commanders (COCOMs), and it facilitates the transfer of funds to DoD components to resolve immediate warfighter needs (IWNs). The JRAC provides COCOMs with the means to rapidly resolve shortfalls and capability gaps identified during ongoing operations (USD[A&S], undated). The JRAC reports to USD(A&S).

The originator recommends a JUON and obtains approval of the general officer in the chain of command. JUONs are submitted to COCOMs for certification and prioritization. The COCOM then rejects or certifies and prioritizes JUONs and submits to the Joint Staff and JRAC simultaneously. With Joint Staff recommendation, the JRAC designates or declines the JUON as an IWN within 14 days of submission to JRAC. The JRAC then tracks the IWN and facilitates its resolution and forwards it to the appropriate DoD Component for action.

Part of the concept behind the JRAC, and the (J)UON and (J)EON process, is that waivers are encouraged whenever possible to streamline the process. The JRAC has $100 million in authority (per FY) to reallocate funding regardless of appropriation category, although Congress must be notified within 15 days of moving money to address a validated need (GAO, 2012a).

Despite these efforts to make (J)UONs the vehicle of choice for addressing IWNs, we found in our interviews that very few agencies were using them as a vehicle for more-responsive acquisition. The process is still bound by JCIDS, and acquiring waivers also requires administrative overhead. The process also does poorly in cases in which it is difficult to define the consequences of not meeting a need (e.g., what happens if a particular cyber capability is lacking). It also is insufficient to address opportunities to gain a new capability that is particularly desirable at a low cost (e.g., giving the SM-6 anti-surface warfare [ASuW] capabilities through a software upgrade).

Additionally, the authority to move money to address a need lies with the Comptrollers rather than with local commanders (as it does with SOF). Identifying potential funding sources and obtaining reprogramming approval is often the most challenging aspect for rapid acquisitions.

Figure 4.1 shows the process for initiating a (J)UON or JEON; validating the requirement; acquiring it; field testing it; then transitioning it to disposal, sustainment, or acquisition as an enduring capability. The process accelerates requirement validation, contracting, fielding, and testing as integrated parts of the process. The process is overseen by a board of directors, which bypasses many of the standard reviews and staff processes under DoDI 5000.02. Up to $100 million per year can be moved to cover these projects, regardless of the appropriation. A full description of this process is contained in AFI 10-601.

Figure 4.1. (J)UON or JEON Process Flow

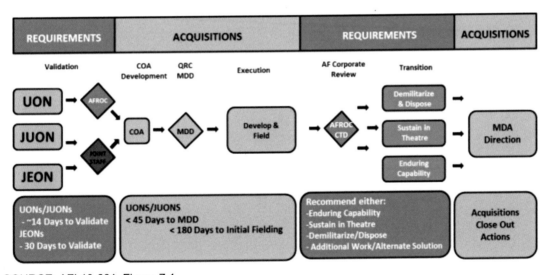

SOURCE: AFI 10-601, Figure 7.1.
NOTE: AF = Air Force; AFROC = Air Force Requirements Oversight Council; COA = course of action; CTD = Capability Transition Decisions; MDD = Materiel Development Decision; QRC = Quick-Reaction Capability.

Table 4.9 lists the individual more-responsive acquisition approaches that are integrated in the JRAC's overall acquisition formula. Table 4.10 briefly outlines how the combinations of these approaches might affect six implementation areas (acquisition strategies, acquisition processes, program structures, headquarters structures, requirements, and budgeting).

Table 4.9. JRAC's Acquisition Formula

Category	Approaches Used	Requirements	Financial	Technical	Acquisition	Intelligence
					Domain	
High level	• Acquisition tailoring				X	
Intelligence	• Faster intelligence				X	X
Requirements	• Requirements: Partial (80%) solutions acceptable to users	X				
Schedule	• Hard, fixed schedules	X			X	
Financial	• Reprogramming		X			
Technology	• Operational prototyping: components					
Reuse	• Modifications to existing platforms	X			X	
	• Repurposing	X			X	
	• Off-the-shelf: COTS or GOTS	X		X	X	
Testing	• Test to acceptance	X			X	
	• Operational testing in actual operations	X			X	
Rapid fielding	• (J)UONs, (J)EONs	X				
Contracting	• Noncompetitive contracting: J&As				X	
	• Use technical integration contract modifications or ECPs				X	
	• Contract schedule incentives				X	
Oversight	• Senior board of directors (direct approval or oversight to PM)	X	X	X	X	
	• Reinforced acquisition chain of command				X	

Table 4.10. JRAC Implementation Considerations

Acquisition Strategies	Uses (J)UONs to address COCOM urgent requirements via accelerated JCIDS process
Acquisition Processes	JRAC collaborates with Joint Staff Gatekeeper and Functional Capabilities Boards in review of proposed JUONs and JEONs prior to validation
Program Structures	Accelerated validation, contracting, fielding, and testing as part of JUON JCIDS process
Headquarters Structures	Use of a board of directors bypasses standard headquarters reviews and staff processes
Requirements	(J)UON process to create validated requirements
Budgeting	$100 million per year to move regardless of appropriation; up to $10 million per project (as is standard for Below Threshold Reprogramming)

Air Force Rapid Development Integration Facility

The USAF Life Cycle Management Center's Rapid Development Integration Facility (RDIF) began operations in 2010. It is intended to rapidly produce prototypes using in-house government capabilities and engineering. This provides the additional benefit of keeping all IP under government control. The RDIF facility is a 20,000-square-foot manufacturing and modification building at Wright-Patterson AFB, Ohio (Farnsworth, 2015).

Since 2010, the RDIF has completed more than 240 projects on equipment such as the HH-60M Black Hawk Helicopter, F-22 Raptor, B-2 Spirit Bomber, B-1 Lancer Bomber, all variations of the C130J, various Federal Aviation Administration aircraft, and the Guardian Angel Air Recovery vehicle. The types of projects for these airframes include a nitrogen purge kit for infrared sensors, engineering improvements maintenance lift stands for the C-5M Galaxy II, and weapon rack test kits for the F-22A (88th Air Base Wing Public Affairs, 2018). The new nitrogen purge system kits cost $398, compared with the prior capability that cost $6,000.

RDIF personnel have reported returning more than $150 million to customers, who could then use that money on additional projects; RDIF has also reduced time on projects and procurement. RDIF capabilities feature designing prototypes, working with the customer to fine-tune them, testing them, and estimating the cost. RDIF also has a limited production capability. Because of the small scale of these projects, they do not fall under usual JCIDs procedures. They are, however, subject to the FAR.

Table 4.11 lists the individual more-responsive acquisition approaches that are integrated in the RDIF's overall acquisition formula. Table 4.12 then briefly outlines how the combinations of these approaches might affect six implementation areas (acquisition strategies, acquisition processes, program structures, headquarters structures, requirements, and budgeting).

Table 4.11. RDIF's Acquisition Formula

Category	Approaches Used	Requirements	Financial	Technical	Acquisition	Intelligence
General	• In-house engineering and production			X	X	
Requirements	• Requirements: Partial (80%) solutions acceptable to users	X				
Schedule	• Hard, fixed schedules	X			X	
Technology	• Operational prototyping: components					
Reuse	• Modifications to existing platforms	X			X	
Testing	• Test to acceptance	X			X	
Rapid fielding	• (J)UONs, (J)EONs	X				
Contracting	• Indefinite-delivery contracts (e.g., IDIQs)				X	

Table 4.12. RDIF Implementation Considerations and Conditions

Acquisition Strategies	Rapid prototyping; does not use (J)UONs or JEONs
Acquisition Processes	In-house engineering teams to produce prototypes onsite
Program Structures	Small, independent teams of engineers; flat chain of command at facility
Headquarters Structures	Part of the Air Force Life Cycle Management Center but does projects for other programs (e.g., F-22A Raptor program)
Requirements	Generally, provides capabilities to meet existing requirements, but does so more quickly and cheaply
Budgeting	Funding for projects comes from other USAF programs; small projects only; limited production capacity

Defense Innovation Unit

DIU (formerly the DIU Experimental [DIUx]) was established after Secretary of Defense Ash Carter's Drell Lecture at Stanford (Carter, 2015). The DIU's basic mission is rapid adoption of COTS technology into the DoD.

DIU opened in September 2015 to accelerate commercial innovation to the warfighter in order to meet the changing demands of modern strategic and technological environments. The DoD's 2018 National Defense Strategy supported the widely held belief that the nation's military-technical advantage is eroding as competitors and adversaries have the same access to the global technology marketplace driving innovation (OSD, 2018b). Headquartered in Mountain View, California, DIU has offices in Austin, Texas; Boston, Massachusetts; and the Pentagon.

DIU asserts that the relationship between technology development and military involvement has changed fundamentally from past decades when new technologies were often developed by government funding, initially for military use (DIUx, 2017, p. 2). By developing new technologies, the DoD was often the *first mover* until the end of the Cold War; now the DoD is seeking to be a *fast follower* that works hard to keep pace with commercial refresh cycles in key technology areas as the

> commercial sector leads the way in many cutting-edge technologies from artificial intelligence to autonomous systems to space, the convergence of which generates the prospect of great changes to the character of warfare. The implications of global access to advanced commercial technology are visible in today's conflicts and the loss of exclusivity means the likelihood of technological surprise is far higher. (DIUx, 2017, p. 2)

DIU's mission is to "lead DoD's break with past paradigms of military-technical advantage to become fast adapters—as opposed to sole developers—of technology, helping to integrate advanced commercial capability for strategic advantage" (DIUx, 2017, p. 2). DIU is prioritizing "speed of delivery, rapid and modular upgrades, and quick operational adaptation on the battlefield" (DIUx, 2017, p. 2). For warfighters, speed to effectiveness is paramount. On average, DIU reports that its Commercial Solutions Openings[18] take 90 calendar days on average from

---

[18] A *Commercial Solutions Opening* is a type of competitive OT solicitation for a prototype (DIU, 2018).

first contact with a company to contract award (DIUx, 2017, p. 4). This rapid approach not only allows the DoD to properly evaluate the performance and cost of new capabilities; it also provides warfighters the opportunity to develop new concepts of operation before committing to a larger purchase.

DIU uses a multistep, co-investment model that solicits input through such commercial-friendly means as the solutions brief (Carter, 2015; DIU, undated). Once a suitable proposal is identified, DIU provides follow-through services, such as using a flexible contracting mechanism to craft an agreement for the government and technology innovators to collaborate within 60 days after contract negotiations begin. DIU might also choose to contribute some of its own funds to an effort. In addition, DIU performs contract administrative functions and monitors the progress of active contracts, thus offering speed, access, and administrative support to the DoD and military departments.

DIU designated five focus areas in 2017: artificial intelligence, autonomy, human systems, information technology (IT), and space. This approach builds subject-matter expertise within DIU while focusing on relationship development and ensuring unity of effort.

In the 2016 NDAA (Pub. L. 114-92, 2015), Congress granted the DoD OT authorities to allow successful prototype projects to serve as justification for follow-on production contracts without the need for further competition. This helps to enable commercial innovation to survive the gap that can separate newer capabilities from programmed capabilities for warfighters. DIU designed these OT production contracts to scale across the DoD; any DoD entity might buy and use successfully prototyped capabilities without having to allocate time and resources into putting a new contract into place. This practice has a multiplier effect and helps technologies get to the warfighter faster, regardless of service.

For example, DIU has identified companies that provide low-cost, precise, and on-demand deployment of small payloads into space, which will potentially reduce the satellite inventory waiting to go into orbit. This capability will enable new microsatellite constellations, coupled with advanced imagery analytics, to improve peacetime indications and warnings.

DIU represents a combination of methodology, geography, network, and experience that positions DIU to coordinate with key organizations across the DoD's innovation ecosystem, such as the SCO, DARPA, the Defense Digital Service, and the Defense Innovation Board.

Looking ahead, DIU is examining how to scale its processes to encompass the understanding and use of the OT authority and the technologies and methodologies available to the DoD. DIU's success to date has been due, in large part, to the commitment of Army Contracting Command–New Jersey.

DIU and the Defense Digital Service are working to apply agile methodologies to ongoing hardware and software upgrades of legacy platforms. DIU will address challenges that MDAPs will face as they refresh and upgrade hardware and software components.

DIU is trying to lead the DoD toward rapidly acquiring commercial innovation that can solve mission-critical problems by harnessing the disruptive capabilities being imagined by today's commercial technology companies. DIU is another DoD pathway to accessing commercial innovation.

Table 4.13 lists the individual more-responsive acquisition approaches that are integrated in the DIU's overall acquisition formula. Table 4.14 then briefly outlines how the combinations of these approaches might affect six implementation areas (acquisition strategies, acquisition processes, program structures, headquarters structures, requirements, and budgeting).

Table 4.13. DIU's Acquisition Formula

Category	Approaches Used	Requirements	Financial	Technical	Acquisition	Intelligence
High level	• Acquisition tailoring				X	
Combined	• "Agile" development	X			X	
General	• Standing FAR waivers	X	X		X	
Requirements	• Requirements: Partial (80%) solutions acceptable to users	X				
Schedule	• Hard, fixed schedules	X			X	
Financial	• Reprogramming		X			
Technology	• Operational prototyping: components					
Reuse	• Off-the-shelf: COTS or GOTS	X		X	X	
Testing	• Test to acceptance	X			X	
	• Operational testing in actual operations	X			X	
Contracting	• Indefinite-delivery contracts (e.g., IDIQs)				X	
	• Noncompetitive contracting: J&As				X	
	• Sole-source production following successful prototypes				X	
	• Custom IP arrangement via an OT			X	X	
	• Team with commercial partner via an OT			X	X	
	• Contracting that can avoid protests to the GAO (e.g., OTs)				X	
	• Contract schedule incentives				X	

Table 4.14. DIU Implementation Considerations and Conditions

Acquisition Strategies	Rapid prototyping with bridging strategy between JCTD and production for smaller elements; often uses IDIQ contracts, OTs, and Broad-Area Announcements
Acquisition Processes	Work as industry consortiums to respond rapidly to emergent requests for proposals (RFPs)
Program Structures	Industry consortiums managed by commercial contractors
Headquarters Structures	No effect
Requirements	Requirements address emerging needs and opportunities; formal JCIDS requirement process bypassed, kept simple and capability based
Budgeting	Focuses on immediate program needs that already have funding

Army Consortium for Command, Control and Communications in Cyberspace

C5 is a consortium of leading companies and institutions in the C4ISR and cybertechnology sectors.[19] C5 differs from DIU mainly in that it involves more of the larger industry players; DIU is more focused on the small-business innovator. However, there are hundreds of partners in the C5 consortium (Consortium Management Group, undated).

OSD requested that the Army establish a new OT to address C4ISR and cybertechnology requirements. In response, C5 was formed. Based on its business model, C5 was awarded an initial three-year OT with the Army in July 2014 and a ten-year, no-ceiling, follow-on OT in April 2017.

Consortiums such as C5 are increasingly popular. As of September 2017, there were 14 DoD OT consortiums, with a fifteenth scheduled to be established soon (Wōden, 2017). Congress amended and further expanded the DoD's OTA in the FY 2016 NDAA (Pub. L. 114-92, §815, 2015) in 10 USC §2371b, adding the ability for the government to more easily transition a successful OT prototype project directly to a follow-on production contract when the initial prototype effort is competitively awarded.

The Army uses the IT Box model for cyber tool development, which might be applicable to satellite command and control architecture development. This could also be thought of as part of the organizational model that makes the Army C5 model attractive for some projects.

This model possesses the following possible advantages:

- It broadens the technology base by reaching innovators not readily available to the DoD.
- It converts government contracting to commercial practices.
- It provides rapid acquisition (projects costing less than $100 million can be competitively awarded quickly).
- It provides easier transition from prototype directly into production.
- It is open to large and small businesses.
- It offers flexibility in the treatment of IP.

---

[19] C4ISR = command, control, communications, computers, intelligence, surveillance, and reconnaissance.

- It has a platform for public and private collaboration.
- Project awards cannot be protested.

Although likely less revolutionary than DARPA, one of the greatest advantages is a more streamlined transition from prototype to production. C5 contracts, such as DIU and DARPA, are frequently OTs, which offer some advantages in reducing the amount of time it takes to get a project under contract. Additionally, with funding already allocated to be put under contract via an OT, there is less of a lag between identifying a need or opportunity and getting it under contract.

The model of IDIQ using OTs as contract vehicles appears to be gaining popularity for rapid acquisition of ACAT II and III projects costing less than $100 million as other services and programs embrace this option.

Table 4.15 lists the individual more-responsive acquisition approaches that are integrated in C5's overall acquisition formula. Table 4.16 then briefly outlines how the combinations of these approaches might affect six implementation areas (acquisition strategies, acquisition processes, program structures, headquarters structures, requirements, and budgeting).

Table 4.15. Army C5's Acquisition Formula

Category	Approaches Used	Requirements	Financial	Technical	Acquisition	Intelligence
Requirements	• Requirements: Partial (80%) solutions acceptable to users	X				
	• Requirements: keep stable (avoid creep)	X				
	• Bypass JCIDS and JROC	X				
Schedule	• Hard, fixed schedules	X			X	
Financial	• Reprogramming		X			
Technology	• Operational prototyping: components					
Testing	• Test to acceptance	X			X	
	• Operational testing in actual operations	X			X	
Contracting	• Noncompetitive contracting: J&As				X	
	• Marketplaces				X	
	• Bids open only to members				X	
	• Custom IP arrangement via an OT			X	X	
	• Team with commercial partner via an OT			X	X	
	• Contracting that can avoid protests to the GAO (e.g., OTs)				X	
	• Proven contractors only				X	

Table 4.16. C5 Implementation Considerations and Conditions

Acquisition Strategies	Rapid prototyping with bridging strategy between JCTD and production for smaller elements; often uses IDIQ contracts, OTs, broad agency announcements
Acquisition Processes	Work as industry consortiums to respond rapidly to emergent RFPs
Program Structures	Industry consortiums managed by commercial contractors
Headquarters Structures	No effect
Requirements	Requirements address emerging needs and opportunities; formal JCIDS requirement process bypassed, kept simple and capability based
Budgeting	Focuses on immediate program needs that already have funding

Defense Advanced Research Projects Agency

DARPA was established in 1958 as a response to the launch of Sputnik I to help ensure that the United States did not fall behind the Soviet Union in technology. DARPA's mission statement is "to make pivotal investments in breakthrough technologies for national security." Its innovations include the Internet; reduced instruction set computer, or RISC, computing; global positioning satellites; stealth technology; unmanned aerial vehicles, or "drones"; and micro-electro-mechanical systems.

DARPA has had success with rapid development of technology with a modest budget and a relatively small staff of about 240 government employees, almost one-half of whom are management. DARPA partners with industry, academia, and other government agencies to pull in a broad array of experience and knowledge.

The organizational model at DARPA consists of three primary ingredients: ambitious goals, temporary project teams, and independence (Dugan and Gabriel, 2013).

DARPA aims to make revolutionary technological leaps rather than evolutionary ones. It brings together world-class experts from industry and academia to work on projects of relatively short duration with concrete, closed-ended goals. Finally, by its own charter, DARPA has autonomy in selecting and running projects, which accelerates the development timeline, reduces administrative overhead, and helps attract top talent to projects.

DARPA bases project selection on (1) identifying emerging user needs that existing technologies cannot address or (2) recognizing that a scientific field has emerged or reached an inflection point and can be used to solve a practical problem of importance, often in a new way. The latter can be poorly suited to a traditional DoD process in that these new technological opportunities might create a new demand that is not currently a validated requirement. DARPA has a much higher risk tolerance than traditional DoD acquisition; it expects some technologies to fail simply as the cost of doing business. Additionally, DARPA will develop technologies that might not have an immediate defense application on the expectation that new technical capabilities will attract requirements.

DARPA does not rely on (J)UONs or (J)EONs to determine which technologies to pursue, but it might be asked to fulfill them. It does not rely on JCIDS as part of the development but does have to comply with the FAR. However, this is where the process breaks down. DARPA develops projects only to the point of being technology demonstrators, at which point the project funding must come from other parts of the department that are subject to the JCIDS process. Lack of funding for technology transitions has always been a significant challenge that DARPA is well aware of and frequently studies. Figure 4.2 illustrates this issue, often referred to as the "Valley of Death." This gap in transition funding from prototypes to operational production can sometimes be avoided by getting early buy-in from potential sponsors, using preexisting requirements, or canceling a program early on if it becomes clear there is not a demand for the prototype. Other support options to bridge the gap are Small-Business Innovation Research and Small-Business Technology Transfer funding and the Rapid Innovation Fund (for example, see DPAP, 2003; DAU, 2005; Hagan, 2011; and DARPA Small Business Programs Office, 2018).

Figure 4.2. Technology Transition Challenge

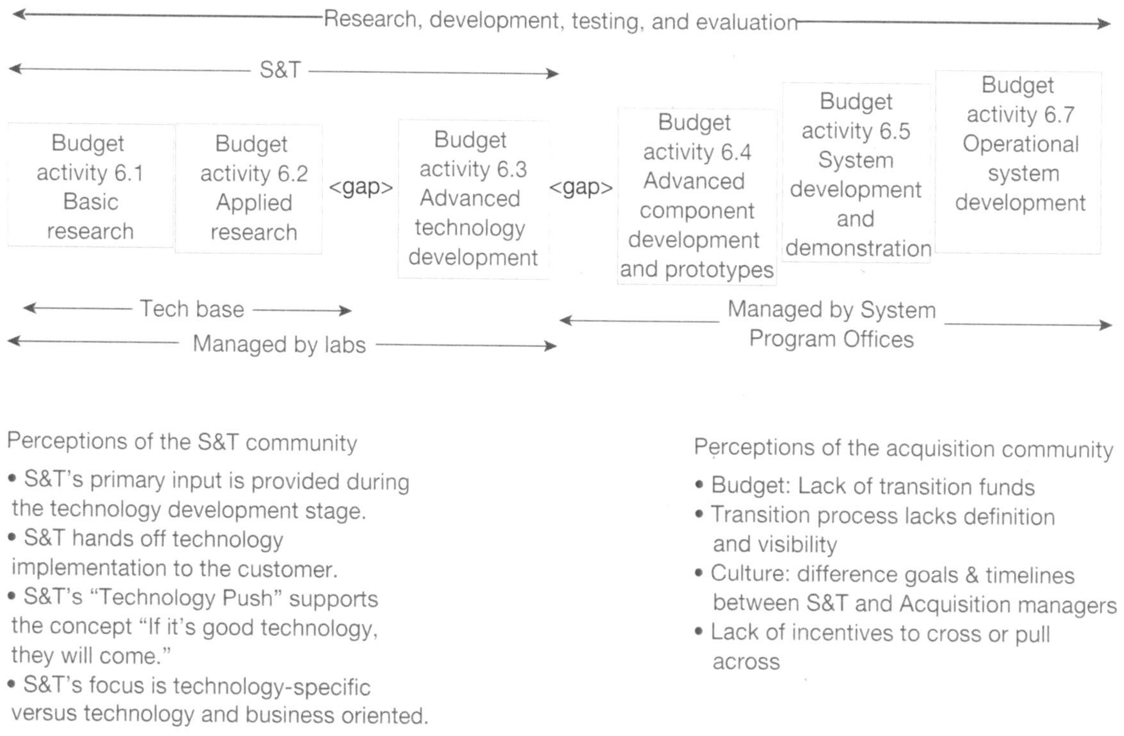

SOURCES: Hagan, 2011; DoD Financial Management Regulation 7000.14-R, 2017.

DARPA has been authorized to use OTs since 1989, predating other DoD authorizations, allowing it to bypass some FAR Part 30 Cost Accounting Standards (FAR Part 30, 2019). One type of OT used by DARPA is the Technology Investment Agreement (TIA); another is an OT for prototypes. TIAs are typically used when the primary goal of the agreement is to perform a research effort, even if items need to be created to test the credibility of the research. TIAs, which are used in accordance with 10 USC §2371 and Part 37 of the DoD Grant and Agreement Regulations (Code of Federal Regulations, 2003), are meant to encourage participation from commercial firms that otherwise might not develop defense technologies (DARPA, undated-a).

OTs for prototypes are used when the main focus of the agreement is to create a prototype. They are used by government to define a relationship with industry, similar to C5 and DIU.

Table 4.17 lists the individual more-responsive acquisition approaches that are integrated in DARPA's overall acquisition formula. Table 4.18 briefly outlines how the combinations of these approaches might affect six implementation areas (acquisition strategies, acquisition processes, program structures, headquarters structures, requirements, and budgeting).

Table 4.17. DARPA's Acquisition Formula

Category	Approaches Used	Requirements	Financial	Technical	Acquisition	Intelligence
High level	• Acquisition tailoring				X	
Requirements	• Bypass JCIDS and JROC	X				
Technology	• Operational prototyping: platforms			X	X	
	• Operational prototyping: components					
	• JCTDs	X		X	X	
	• Prize competitions			X		
Reuse	• Modifications to existing platforms	X			X	
	• Adaptation of foreign technology			X	X	
Testing	• Test to acceptance	X			X	
	• Operational testing in actual operations	X			X	
Rapid fielding	• (J)UONs, (J)EONs	X				
Contracting	• Custom IP arrangement via an OT				X	X
	• Team with commercial partner via an OT				X	X

Table 4.18. DARPA Implementation Considerations and Conditions

Acquisition Strategies	Produce technologies and JCTDs as prototypes, leaving full acquisition to the Components (if pursued further); often uses OTs (TIAs and prototypes) and broad agency announcements
Acquisition Processes	Ad hoc teams of academics and industry to achieve short-term goals; addresses emerging needs and applying new technologies to those needs
Program Structures	Works with potential programs of record and the requirements community to facilitate eventual transfer of the technology; programs should be identified and structured to receive the results should they be successful
Headquarters Structures	S&T organization reports directly to OSD
Requirements	No effect (often just general needs at this point or technology push before formal requirements)
Budgeting	Some flexibility to address technology opportunities and JCTDs; results struggle with transition funding

Lockheed Martin's Skunk Works (and Similar Concepts)

Lockheed-Martin Skunk Works is the official pseudonym for the Lockheed Advanced Development Program. It was formed in 1943 by Kelly Johnson to build a jet fighter around the British "Goblin" engine that could match the German ME 262 jet fighter. This then-classified U.S. project eventually became the P-80A Shooting Star. During the Cold War, Skunk Works produced many (now) famous advanced technology aircraft, such as the SR-71, U-2, and F-117 Nighthawk (Miller, 1995).

Because of these successes, "Skunk Works" has become synonymous in industry with secret projects that use small, loosely structured teams of highly skilled engineers and innovators to develop radical technological breakthroughs. Boeing's variant is called Phantom Works. Google and Apple have both used the Skunk Works concept to develop project offices dedicated to secret projects aimed at developing revolutionary new technologies.

The philosophy of Skunk Works can be summed up as small teams with the best people. Kelly Johnson, who was the first Skunk Works team leader, summed up his philosophy in 14 rules:

1. The Skunk Works manager must be delegated practically complete control of his program in all aspects. He should report to a division president or higher.

2. Strong but small project offices must be provided both by the military and industry.

3. The number of people having any connection with the project must be restricted in an almost vicious manner. Use a small number of good people (10% to 25% compared to the so-called normal systems).

4. A very simple drawing and drawing release system with great flexibility for making changes must be provided.

5. There must be a minimum number of reports required, but important work must be recorded thoroughly.

6. There must be a monthly cost review covering not only what has been spent and committed but also projected costs to the conclusion of the program.

7. The contractor must be delegated and must assume more than normal responsibility to get good vendor bids for subcontract on the project. Commercial bid procedures are very often better than military ones.

8. The inspection system as currently used by the Skunk Works, which has been approved by both the Air Force and Navy, meets the intent of existing military requirements and should be used on new projects. Push more basic inspection responsibility back to subcontractors and vendors. Don't duplicate so much inspection.

9. The contractor must be delegated the authority to test his final product in flight. He can and must test it in the initial stages. If he doesn't, he rapidly loses his competency to design other vehicles.

10. The specifications applying to the hardware must be agreed to well in advance of contracting. The Skunk Works practice of having a specification

section stating clearly which important military specification items will not knowingly be complied with and reasons therefore is highly recommended.

11. Funding a program must be timely so that the contractor doesn't have to keep running to the bank to support government projects.

12. There must be mutual trust between the military project organization and the contractor, the very close cooperation and liaison on a day-to-day basis. This cuts down misunderstanding and correspondence to an absolute minimum.

13. Access by outsiders to the project and its personnel must be strictly controlled by appropriate security measures.

14. Because only a few people will be used in engineering and most other areas, ways must be provided to reward good performance by pay not based on the number of personnel supervised. The Skunk Works manager must be delegated practically complete control of his program in all aspects. He should report to a division president or higher. (Lockheed Martin, undated)

Skunk Works is an industry organization and can be part of any available contracting or acquisition process—MTA, for instance, which also bypasses JCIDS. However, the size of the projects (often below ACAT I) and their secrecy somewhat reduce oversight requirements. Waiver requests are a frequent part of development. The chain of command in a project is kept fairly flat relative to the small size of the teams and decision authority being pushed to the lowest level possible.

Kelly once remarked that, "You can't stack enough average people high enough to equal one good person" (National Research Council, 2011). People assigned to Skunk Works are regarded as part of the corporate "A Team," whether as designers or contracting officers. This means that the model might not be feasible under all circumstances because not everyone can be above average.

However, many of Kelly's rules are applicable across the DoD for more-responsive acquisition, such as the admonitions to keep funding streams steady, use COTS when possible, and build mutual trust between the government and the contractor.

Table 4.19 lists the individual more-responsive acquisition approaches that are integrated in the Skunk Works overall acquisition formula. Table 4.20 then briefly outlines how the combinations of these approaches might affect six implementation areas (acquisition strategies, acquisition processes, program structures, headquarters structures, requirements, and budgeting).

Table 4.19. Skunk Works–Like Acquisition Formula

Category	Approaches Used	Requirements	Financial	Technical	Acquisition	Intelligence
Combined	• Skunk Works–like organizations	X	X	X	X	
Schedule	• Crashing the schedule		X	X	X	
	• Fast-tracking (parallelization; concurrency)				X	
	• Streamlining				X	
	• Focused work				X	
	• Work-breakdown schedule reviews				X	
Technology	• Operational prototyping: platforms			X	X	
	• Operational prototyping: components					
Reuse	• Modifications to existing platforms	X			X	
Contracting	• Proven contractors only				X	
Oversight	• Reinforced acquisition chain of command				X	

Table 4.20. Skunk Works–Like Organizations Implementation Considerations and Conditions

Acquisition Strategies	COTS, GOTS if possible; rapid point solutions; partial solutions acceptable
Acquisition Processes	Follows any applicable acquisition process or contracting mechanism given the nature of the activity, including DoDI 5000.02 and FAR; waivers encouraged; warfighter involvement at all phases; features "A Team" players, including top-flight contracting officers
Program Structures	Short chain of command; small IPTs, such as legal, engineering, financial, and contracting; experienced PMs given high degree of freedom and control
Headquarters Structures	No effect
Requirements	Response to COCOM urgent requirements; (J)UON, JEON process
Budgeting	Follows FAR and DFARS; enhanced ability to move DAF funding to the DAF RCO; attempts to maintain steady funding streams

Strategic Capabilities Office

The SCO was established in 2012 by then–Deputy Secretary of Defense Ash Carter to take military systems that do one thing and make them do something completely different. The goal was to maintain a good balance between the capabilities that the United States shows to the world for deterrence versus those that it keeps secret for warfighting overmatch. An example of this is the funding that the SCO provided to develop SM-6 ASuW over-the-horizon targeting capability to act as a deterrent by repurposing the existing surface-to-air missile for surface-to-surface missions. The SCO uses OTs to rapidly acquire hardware for its demonstration programs.

An SCO program demonstrated a potential disruptive capability in the Perdix system using a series of small drones with COTS technology that, after launch, are able to self-organize into a swarm of small ISR devices. Future versions of this could be launched from the flare dispensers of an F-16 or F/A-18 fighter jet.

In 2018, the office was focused on such artificial intelligence projects as Palladium, which is a broader Navy logistics effort that involves smart sustainment in support of fourth-generation fighter aircraft. A subproject for that is Jarvis, which involves putting a robotics suite into the field with maintainers that can scan existing parts and quickly remanufacture them (Mehta, 2018).

These capabilities have already been demonstrated, but it remains to be seen whether these systems will be integrated into the force over the long term. The exception to this is SM-6. The system was in development as a surface-to-air system. An ASuW capability was added via a software modification, and this ASuW capability is now part of the Navy's long-term CONOPS for the SM-6.

The SCO now reports to the Deputy Secretary of Defense (Pub. L. 116-92, 2019), but its future is uncertain. Although the SCO's budget increased by 23 percent to $1.39 billion in 2019, the 2019 NDAA directed the Secretary of Defense, acting through the USD(R&E), to submit a plan by March 1, 2019, on whether to eliminate, transfer, or retain the SCO (Pub. L. 115-232, §217). As of early June 2019, we had not seen the plan.

Table 4.21 lists the individual more-responsive acquisition approaches that are integrated in the SCO's overall acquisition formula. Table 4.22 then briefly outlines how the combinations of these approaches might affect six implementation areas (acquisition strategies, acquisition processes, program structures, headquarters structures, requirements, and budgeting).

Table 4.21. SCO's Acquisition Formula

Category	Approaches Used	Requirements	Financial	Technical	Acquisition	Intelligence
High level	• Acquisition tailoring				X	
Technology	• Operational prototyping: platforms			X	X	
	• Operational prototyping: components					
Reuse	• Off-the-shelf: COTS or GOTS	X		X	X	

Table 4.22. SCO Implementation Considerations and Conditions

Acquisition Strategies	Modification or repurposing of COTS or GOTS systems
Acquisition Processes	Higher risk tolerance; OTAs, technical integration contract modifications, or ECPs
Program Structures	Specialize programs to facilitate development in close coordination with sponsor
Headquarters Structures	SCO now reports to the USD(R&E); no information on any considerations for internal structure, but reporting to a USD instead of the Secretary of Defense means the SCO has less political clout inside the DoD
Requirements	Does not use validated or JCIDS requirements, but addresses perceived warfighter needs; often just general at this point or a technology push before formal requirements
Budgeting	Flexibility to address technology opportunities; results struggle with the transition funding gaps

Organization Model Observations and Commonalities

One of the overarching themes of the organizational case studies is that the process tends to fit the need. The NRO uses engineering approaches to gain additional capability from satellites that are already in orbit when there is an immediate need. USSOCOM uses COTS and flexible contracting to acquire capabilities in a similar time frame. There are several different models for rapid prototyping, such as DARPA, RDIF, Big Safari, C5, and DIU, that vary depending on what each organization is trying to produce. For even larger, longer-term programs, the DAF RCO and Skunk Works share some similar approaches but also differ—the DAF RCO being a government entity and Skunk Works being a defense industry entity. However, certain themes emerged along the six categories of accelerated acquisition examined.

Acquisition strategies observed in these organizations tend to push decisionmaking to the lowest level possible. These organizations also try to avoid "blank slate" approaches, relying on incremental development, COTS, and GOTS when possible (though DARPA is an exception). Most programs try to mitigate JCIDS (with waivers), tailor it to the maximum extent possible, or avoid it altogether.

Acquisition processes used by these organizations tend toward small IPTs and "A Teams" on critical projects. There is a much higher risk tolerance for many of them, particularly at DARPA. The organizations working with contractors tend to have higher expectations and try to maintain greater degrees of accountability. Such organizations as Skunk Works, the DAF RCO, and USSOCOM try to have a flat chain of command to reduce the amount of administrative and bureaucratic overhead.

Program structures or organizations are tailored to the individual program, what it is trying to produce, its acquisition strategy, and the operational processes it is working under. For example, a program that has an ability to seek tailoring of requirements might need staff and organizations that interact with users and the formal requirements community to update them on the cost, schedule, technical maturity, and risks during development. On the other hand, a

108

program that is expected to meet the requirements given to it regardless of difficulties might focus resources more on technical development and managing requirements with the contractor and have less structural ability to interface with users and the formal requirements community. Such organizations as Skunk Works, DAF RCO, and USSOCOM put as much authority and control as possible at the local program level, so their program structures will need to feature elements to strategize, manage, and control challenges as they arise. Others, such as Big Safari and USSOCOM, focus on rapid COTS or GOTS acquisition and insertion; this means that their structures and staff need to feature such elements as readily available engineering design capabilities and rapid market research to aid in quickly identifying mature technologies; adjusting designs accordingly; and moving out smartly with contracts, development, and production.

Headquarters structures are components in a few of the organizations surveyed. Concepts for new capabilities or technological concepts often need senior leadership support. The use of flat chains of command is meant to avoid layers of staff review.

Requirements management was a component of several of these organizations' overall strategies. Often, expedited JCIDS validation of requirements or direct support from a Military Service Chief is necessary. Some organizations—such as DARPA, C5, and DIU—skip the need for validated requirements entirely, relying on input from the field or recognition of technological opportunities. Additionally, flexibility in requirements allows the product to meet the latest needs as the battlespace evolves. Another common strategy is acceptance of 80-percent solutions to get useful capability fielded quickly, even if all performance metrics have not been achieved. Yet another approach is to set high-level capability-based descriptions of requirements and minimize the use of numerous and rigid key performance parameters and key system attributes.

Budgeting strategies are a key part of the accelerated acquisition approach in most of the rapid programs observed. It is important to have the ability to quickly fund against new threats or technology. Some methods to quickly fund acquisition of critical capabilities are the use of funding lines for existing system upgrades, senior leadership support for reprogramming, and POM revisions. The length of POM cycles and reliance on JCIDS can delay the acquisition of crucial technologies and capabilities by years. Funding stability is also important; abandoning promising technologies and capabilities can have unanticipated adverse consequences later.

5. Discussion and Conclusions

Organizational strategies for acquisition agility draw on multiple approaches that address barriers and increase responsiveness across multiple domains and disciplines. The existence of these approaches (including many that are not new), together with the existence of many rapid-acquisition organizations, illustrates that the acquisition community indeed knows how to go faster if speed is of the essence. Further analysis of these approaches identified key conditions necessary for their implementation. The variety of these conditions and the diversity of approaches complicate the process of sorting through which approaches might be appropriate for a given situation, so we developed a simple lookup software tool that extracts potentially appropriate approaches when provided a set of conditions. It should be noted that although this approach produces candidates based on various considerations, the ultimate choice depends on judgments by decisionmakers.

Perspectives on Enablers for Acquisition Agility

As discussed earlier, a common method to speed acquisition is to prioritize schedule over cost or technical performance (i.e., to recognize the so-called iron triangle). Another common view is that acquisition *processes* are the culprit, and, if they could just be reformed, then acquisition would be faster. Our analysis of more-responsive acquisition approaches and organizations found that although there is some truth in these perspectives, the answer is richer.

We sought some insight into what is needed to enable acquisition agility by using the groupings in Box 3.1 of the necessary conditions for the acceleration approaches. We then counted the number of times that each category of condition is needed across the acceleration approaches according to the mapping illustrated in Figure 3.1 between conditions and acceleration approaches (i.e., counting the "X"s). This generally works, except for three technology conditions and two requirements conditions that are actually challenges rather than enablers; they were removed from this counting.[20] This exercise gives a rough sense of both the range and frequency of enablers for acquisition agility (at least as exhibited by the approaches in our report).

Figure 5.1 shows the results. Processes (including contracting) play a part, but other factors, such as requirements (e.g., flexibility and accepting more readily available capabilities) and technology aspects, occur more often. Financial considerations (e.g., mechanisms to obtain budgets) come up about as often as contracting conditions. Workforce capabilities are also important, as are product conditions (generally the use or modification of existing products).

---

[20] One could also address such an exercise by counting the number of acceleration approaches within each category in Box 2.1, but we observed that this would miss the important enabler of workforce.

Low-risk conditions are also helpful (but, of course, might not be viable given the needs). Finally, there were a few instances of using different providers (e.g., government development or strategic partnerships). Again, this is not meant to be a strongly quantitative result but rather to illustrate that enabling acquisition agility requires more than just isolated changes, such as streamlined processes. Multiple enablers are often needed, and they cut across requirements, technology, financial, process, workforce, and other domains.

Figure 5.1. Frequency of Conditions for Responsive Acquisition Approaches

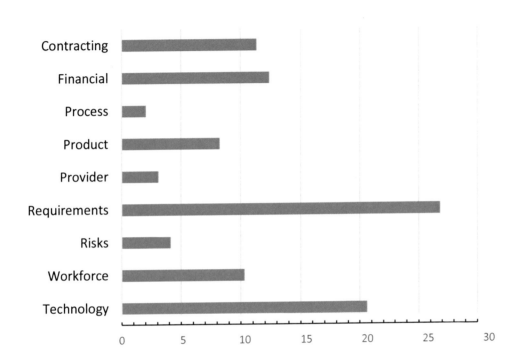

Gaps and Challenges

As the diverse set of acceleration approaches in use makes evident, the acquisition community has ways to go faster, but it cannot do so alone. Fundamentally, acquisition requires three things to proceed: a requirement (articulated and validated in some form), money, and applicable technology or services. Requirements and financial resources are managed by other communities in the DoD, so those also need to proceed quickly or be tailored to readily available technology or services if acquisition is to go faster. Obtaining financial resources on short notice is perhaps the most difficult task because the budget process is conducted annually with a long lead time and Congress ultimately controls not only levels of funding but also what that funding can be applied to. The DoD has long recognized the need for short-term, flexible financial resources to fill the gap, but Congress has been reluctant to allocate moderate flexible resources, even with contemporaneous explanations of what the resources were applied to. Budgetary

111

reprogramming can reallocate funds for critical needs, but thresholds are relatively modest and there is an overall annual limit.

In addition to few budgetary solutions, a simple examination of acceleration approaches by basic domains revealed very few approaches that improve the interface between the acquisition and intelligence communities to better understand and track threat evolution (when systems must stay current against threats).[21] One would hope and expect that the requirements community would serve this function, but requirements are often static. Acceleration approaches that involve working closely with users and the requirements community to evolve requirements as designs and technology are explored during acquisition should help to bridge this gap. Also, further analysis might be warranted in terms of what the acquisition community needs from the Intelligence Community and how these things can best be obtained. It should be noted, however, that evolving requirements can both cause and prevent delays, depending on use.[22]

Conclusions

Acquisition can be rapid. Speed often entails some level of compromise in cost or technical performance objectives, but it also involves other factors, such as investments in workforce expertise and experience, ready availability of financial resources, and a willingness to accept operational capabilities incrementally. Process improvements can help, but without addressing these other factors, the gains are likely to be modest.

The DoD has used various acquisition approaches successfully for decades to improve responsiveness. As threats and technological opportunities increase, the availability of a wide variety of approaches as outlined in this report holds promise that the DoD can keep up with or even stay ahead of threats by leveraging opportunities to retain defensive advantage during operations and by drawing from the variety of supporting functions and capabilities.

---

[21] Note that updating requirements given evolving threats can slow a program relative to the original requirements and schedule and thus be potentially criticized for requirements creep. However, if the program is expected to stay current against evolving threats, then schedule growth will be reduced if the program obtains intelligence on evolving threats sooner rather than later.

[22] Evolving requirements can be used to prevent delays (e.g., deciding that solutions developed so far are good enough for the users) or slow them down through requirements creep.

Appendix A. Additional Insights into Implementation Considerations

Selecting appropriate approaches is one thing. Living with the practical implementations they impose is another. In this appendix, we provide some in-depth examples of how three acceleration approaches might affect six major implementation areas: acquisition strategies, acquisition processes, program structures, headquarters structures, requirements, and budgeting. These practical considerations factor in which of the six areas must deal with the necessary conditions or apply the acceleration approaches involved, along with any additional factors that should be considered beyond those identified in the conditions lists.

We summarize these practical considerations for three example approaches in common use today: MOSA, agile development, and rapid prototyping. Our overview of these three approaches (which are generally applicable to space programs) are largely based on the reviews by Kim et al. (forthcoming) and Camm et al. (forthcoming). More-concise implementation considerations for the rest of the acceleration approaches listed in Box 2.1 are discussed in Chapter 2. Additional discussion specific to the LTPSI approach is in Appendix B. We also provide some exploratory ideas on how to protect S&T maturation budgets in Appendix C.

Modular Open-Systems Architectures

Drawing largely from Kim et al. (forthcoming), Camm et al. (forthcoming), and our personal expertise, we found that MOSA are characterized by consensus and use of widely supported (often commercial) interface standards. The MOSA approach could be applied across the Space Enterprise Vision and has relevance to establishing a common ground-control architecture (i.e., a family of common buses and common bus-to-payload interfaces). With the relatively small quantity of systems produced in support of the Space Enterprise Vision, economies of scale probably can be achieved if commonality can be enforced for systems with similar design requirements and constraints.[23] MOSA also enable enhanced competition, incorporation of innovation, cost savings, interoperability, and a facilitation of technology refreshes. Implementation of MOSA, however, has broad considerations for acquisition programs, processes, and structures.

---

[23] AFSPC is leading an effort to build a Space Force that will prevail in the face of the threat. Specifically, AFSPC envisions an integrated approach—the Space Enterprise Vision—to building a force across all space mission areas, coupling the delivery of space capabilities to the warfighter with the ability to protect and defend space capabilities against emerging threats. The Space Enterprise Vision requires significant changes in how the Air Force designs, develops, acquires, and operates its space systems to build a future space enterprise.

Acquisition Strategies

Acquisition strategies should address systems integration considerations early on. Using a contractor as the system integrator under a MOSA approach would avoid proprietary data challenges between competing component contractors and would require knowledge only of the system design and interface standards. The life-cycle program plan development and systems engineering efforts should be flexible to fully support technology development timelines and should support the system integration effort identified in the strategy.

Acquisition Processes

The acquisition process could focus on incremental development and fielding to deploy the latest sensors and capabilities that conform to the standardized interface and respond to emerging threats. Depending on the system integration strategy and the specific system, some detailed design engagement between component suppliers and system integrators might be required. Rather than the development driving a delivery date, development schedules could be driven by crew training needs, operational needs, and maintenance schedules. These processes need to be flexible enough to accept requirements changes for the good of the enterprise.

Program Structures

Programs need sufficient systems engineering to design the architecture and select interface standards. This needs to happen at both the government and contractor levels.

Headquarters Structures

MOSA require broad, enterprise-wide coordination and might require suboptimal program-level decisions for portfolio- or enterprise-level success. This might require an enterprise-level governance structure to ensure that program-level decisions do not suboptimize the enterprise. This enterprise-level governance would need goals and outcomes closely connected with program-level goals and incentives, sufficient data for informed decisions, senior leadership sponsorship, and a formal structure with well-defined, repeatable process and performance metrics. Another key role of the enterprise-level governance would be to decide which systems would benefit from standardization.

Requirements

The MOSA structure could result in additional requirements not relevant to some specific programs and could create suboptimal solutions in others, which might create problems in the requirements process if PMs are not allowed to compromise on program specifics for broader portfolio or enterprise benefits. If incrementally focused, the requirements process in general would need to be flexible and set thresholds or goals that can change over time to allow for expedient technology insertions.

Budgeting

The budget should reflect any needs to respond to emerging threats and technology developments through incremental upgrades (e.g., to rapidly operationalize new sensors or capabilities to benefit from the standardized bus-to-payload interface). MOSA should enable future upgrades at a lower cost through ease of integration and increased competition for components.

Table A.1 provides a summary of the conditions for implementation of MOSA.

Table A.1. MOSA Implementation Considerations and Conditions

Acquisition Strategies	Relies on open standards to enable more flexible trading of components from multiple vendors
Acquisition Processes	Focus on incremental development and fielding
Program Structures	Needs sufficient systems engineering to design the architecture and select interface standards
Headquarters Structures	Broad, enterprise-wide coordination; might need enterprise-level governance structure
Requirements	Need flexibility; thresholds or goals should be able to change over time to allow for expedient technology insertions
Budgeting	Reflect any needs to respond to emerging threats and technology developments through incremental upgrades.[a]
[a] MOSA programs might need either MRs or reprogramming support; otherwise, the budgetary planning, programming, authorization, and appropriation processes might not respond fast enough to fund the needed changes.	

"Agile" Development

Drawing largely from Kim et al. (forthcoming), Camm et al. (forthcoming), and our personal expertise, we found that agile development is characterized by incremental development and rapid testing within and after each build ("sprint") to respond quickly to technical opportunities and feedback on capability needs. The "Agile" development approach is particularly relevant when future capability needs are unclear or when the environment in which a system is designed to operate becomes increasingly dynamic. New space threats, technology failures, rapid technology developments, and increasing reliance on space technologies make "Agile" development an appealing approach for quick and robust response. Space software and ground control stations are particularly attractive options for the application of "Agile" development, given the ability to apply incremental code changes or upgrades, respectively (Kim et al., forthcoming).

Acquisition Strategies

It is important to understand the overall acquisition setting (e.g., factors and conditions in Box 3.1 and other characteristics of the acquisition situation at hand) to determine the degree, if any, to which an "Agile" development approach should be used. The acquisition strategy will have to recommend whether a traditional approach, "Agile" development approach, or hybrid of the two is best suited to achieve outcomes.

Given the testing necessitated by an "Agile" development approach, the strategy should consider a potential developer's test processes as part of any source selection and coordinate formal DAF testing and in-process testing. The RFP will have to communicate how the program intends to balance skills, experiences, and costs. RFPs should contain discussion of how to coordinate scaling frameworks between the DAF and developers; explicit protocols for managing refactoring to overcome the challenges that emerge from multiteam, parallel efforts; and a method for assessment of progress that allows for flexible forward movement. Source selection should also consider the level of experience and skills a contractor offers and understand that "Agile" development teams will likely need higher levels of expertise both individually and in teamwork, resulting in higher absolute costs. The contracts should also be flexible and incentivize behavior that supports "Agile" development. A cost-based IDIQ contract can support rapid adjustments in content and level of effort.

Acquisition Processes

The "Agile" development process is incremental, deploying and operationalizing systems as early as when they meet minimum requirement thresholds. The process should also allow feedback from users following each iteration to advise future development iterations and should support sprints and releases. A *sprint* consists of several stages from initial planning up to a functional demonstration to users and stakeholders, followed by a release. The sprint-release process is iterated until the program is completed, defining the "Agile" development effort. Developmental testing is incremental and continuous. Operational testing needs to leverage developmental testing, accommodate modular testing of incremental changes, and reflect operational capabilities as viewed by users.

Program Structures

Programs should be structured to support the degree of "Agile" development identified in the acquisition strategy and should integrate users into the acquisition and iteration process to capitalize on feedback for future development. In highly "Agile" development programs, users should be accessible to development and contracting specialists. Program structures, such as cross-functional teams and IPTs, can support agile development and are discussed in greater detail in Chapter 4.

Headquarters Structures

According to the GAO (2012b) and Kim et al. (forthcoming), "Agile" development could benefit from having the "Agile" development implementation monitored by an O-6 (or above) champion who has both experience and a reputation within the requirements, programming, acquisition, and operational communities. For space acquisitions, this champion logically would be at AFSPC. An additional advocate with modern "Agile" software development and acquisition skills who advises policymakers (e.g., throughout AFSPC and SMC for space acquisitions) to support "Agile" development would also benefit implementation. Further analysis would be needed to determine where such a champion and advocate might best reside and whether any major headquarters structures would be affected. Headquarters functions, though, would need to be structured to ensure that acquisition personnel at the buying command and in the program office have training and mentoring in "Agile" best practices beyond what DAU offers in its traditional acquisition certification courses. Additionally, users and requirements developers must be involved in similar training and education and should be involved in the acquisition process. AFSPC will also have to determine which users will be involved in development activities and testing.

Requirements

Requirements must have two things to support "Agile" development. The first is that requirements be framed as a capability to accomplish a mission rather than as a specific engineering design solution aimed at a given mission. The second is a mechanism for identifying the minimum requirement levels that will be valuable to a user. This gives the designer flexibility and allows for capability-driven design while allowing for early user feedback and subsequent incremental improvements.

Budgeting

The program's budget should align with the incremental aspect of development and testing. As the program progresses, the user should judge whether future capability improvements are either mission-relevant or necessary. The user should judge whether mission-relevant capability improvements justify cost increases within each iteration.

Table A.2. "Agile" Development Implementation Considerations and Conditions

Acquisition Strategies	Needs flexible contracts that focus on capabilities delivered during development rather than on fixed product requirements; end state is determined through user feedback on incremental builds
Acquisition Processes	Support sprints and releases; testing is incremental and continuous
Program Structures	Support the degree of "Agile" development; integrates users for feedback
Headquarters Structures	An "Agile" champion and advocate at AFSPC could provide strategic support and facilitate the appropriate use of "Agile" development, including workforce training and empowerment; some leadership support and guidance might be needed
Requirements	Requirements are capability-centric, reflect mission objectives, and have identified the minimum necessary for accomplishing design intent
Budgeting	Budgets cover multiple increments, delivering the best capabilities possible within available funding

Rapid Prototyping

Rapid prototyping is characterized by the ability (sometimes outside a program of record) to explore and develop capabilities quickly. Prototyping can inform acquisition decisionmaking and reduce technical risk through a specific set of design and development activities. Rapid prototyping might be particularly useful in (1) responding to technological surprises that threaten U.S. space systems, (2) implementing planned improvements to systems, and (3) demonstrating and evaluating new technologies operationally (Kim et al., forthcoming).

Acquisition Strategies

For rapid prototyping to be successful, the acquisition strategy should clearly describe the technology development pathway and transition plan and provide measurable and achievable exit criteria. Assuming that the prototype shows achievable utility and is produced instead of disposed of, the strategy should have a transition plan for either immediate operational deployment or transition into a program of record. If the intention is for the prototype to transition to a program of record, the transition plan should consider when in the program (Milestone A, B, etc.) it will transition. Attention should be paid to major program of record events and characteristics, but generally the prototype should be associated and integrated into the program of record as early as possible.

The acquisition strategy for a prototyping activity should also identify the customer and develop an integration plan for adopting and integrating the technology into the customer's community. Early consideration for technical maturity; manufacturing feasibility; sustainment; and tactics, techniques, procedures, and training requirements also improve the likelihood of successful prototype transition and integration. The strategy can also benefit by leveraging recent NDAA provisions related to rapid prototyping (e.g., Sections 804 and 806).

Acquisition Processes

The prototyping phase needs to be reflected in the process. Programs should have processes that support prototyping projects with realistic schedule and cost projections and reasonably mature technologies. Projects should also receive early user endorsement or support and regular communication.

Program Structures

The program structure needs to provide capacity for prototyping design, management, and operational engagement.

Headquarters Structures

Rapid prototyping probably does not affect headquarters structures per se, but extensive operational engagements (e.g., system use during operations or utilization of operational capabilities and platforms for testing) might require approval by headquarters command elements or MDA, ensuring that safety and operational effects are minimal and acceptable.

Requirements

These are unaffected. Validated requirements are not needed until an actual procurement of prototype units in significant numbers takes place. Some insight into operational needs, however, can be useful for guiding prototype R&D. Examples of prototypes that went into low-rate initial production are Predator and Global Hawk.

Budgeting

Resources for prototyping need to be factored into budgets. Funding should be secured early to conduct prototype activities without delay. If prototypes are left for operational use, then sustainment funding will need to be identified to deploy, operate, and sustain fieldable prototypes immediately on completion of development and test activities.

Table A.3. Rapid Prototyping Implementation Considerations and Conditions

Acquisition Strategies	Describe the technology development pathway and technology transition plan with measurable and achievable exit criteria
Acquisition Processes	Reflects prototyping phase in the process plan
Program Structures	Capacity for prototyping design, management, and operational engagement (e.g., system use during operations or utilization of operational capabilities and platforms for testing)
Headquarters Structures	Unaffected, although extensive operational engagements might require approval by headquarters command elements or MDA (e.g., to ensure that safety and mission are assured in any operational prototyping)
Requirements	Validated requirements not needed until an actual procurement of prototype units in significant numbers has occurred; some insight into operational needs might be useful for guiding prototype R&D
Budgeting	Sustainment funding will need to be identified if prototypes left for operational use

Appendix B. Practical Considerations of Keeping the Same Prime System Integrator for the Long Term

One concept raised internally by the DAF for speeding space acquisition was to keep prime system integrators for future major production lots and modifications as long as the bus remains the same, contractor performance is acceptable, and no other major considerations arise. We assessed this LTPSI approach by examining the regulatory forces driving competition, the recent case of the Next-Generation (NextGen) Overhead Persistent Infrared (OPIR) satellite acquisition, factors that drive the competition decision, and how often the DoD rebids the work for competition once an initial contractor selection has taken place on an MDAP.

Note that other acquisition approaches, such as MOSA, could facilitate the LTPSI approach. For example, standardized interfaces in MOSA would facilitate not only efficient upgrades of the system by the LTPSI but also integration of (potentially competitively sourced) components in the system as it evolves. The LTPSI would have the expertise and experience with the MOSA, but the technological details within components could be outsourced. That might facilitate the J&A for the LTPSI.

FAR Justifications for Sole-Source Contracting

In the defense acquisition environment, full and open competition is the prevailing acquisition strategy. This presents a challenge in pursuing an LTPSI approach. Theoretically, the benefits of an LTPSI can exist passively through full and open competition, if the incumbent continues to win the competition by being either the best value offeror or the only offeror. However, if the DAF desires to actively pursue and guarantee an LTPSI approach, it will require something other than full and open competition (i.e., sole-sourcing) after the initial competition for the LTPSI. To go this route successfully, the DAF must document and provide justification that passes the checklist of the FAR. The checklist, captured in FAR 6.302 (2019) explicitly outlines seven circumstances in which strategies other than full and open competition are permitted (Table B.1).

Table B.1. FAR Exceptions for Competition

FAR Paragraph	Circumstances Permitting Other than Full and Open Competition
6.302-1	Only one responsible source and no other supplies or services will satisfy agency requirements
6.302-2	Unusual and compelling urgency
6.302-3	Industrial mobilization; engineering, developmental, or research capability; or expert services
6.302-4	International agreement
6.302-5	Authorized or required by statute
6.302-6	National security
6.302-7	Public interest

SOURCE: FAR 6.302, 2019.

The question is: Where does the LTPSI approach fit? Although the LTPSI language is not used explicitly as a justification in any FAR language, there might be FAR justifications that have intent similar to that of LTPSI. We looked at the recent example of the DAF's decision to sole-source the NextGen OPIR satellite acquisition for clues. Although the DAF did not explicitly use LTPSI as a justification, the justification that was used bears some resemblance to LTPSI.

Example: The NextGen OPIR J&A

The NextGen OPIR satellite acquisition is the follow-on missile warning constellation to the Space Based Infrared System constellation.[24] NextGen OPIR will consist of five space vehicles (SVs) in geosynchronous (GEO) orbit and two SVs in polar orbit. On August 14, 2018, the DAF sole-sourced a $2.9 billion-dollar contract to LMS for three of the five GEO SVs. In compliance with U.S. law, the J&A document supporting the sole-source justification was posted on FedBizOps within 14 to 30 days of contract award (but was redacted to protect proprietary contractor data) (AFSPC, 2018). We reviewed this document and found that the DAF used FAR 6.302-1 (2019) to justify sole-sourcing to the incumbent, LMS (AFSPC, 2018). The FAR condition of "only one responsible source" is broken down into three separate and distinct circumstances. Thus, the J&A must address and provide the rationale for one of the three. We found that the DAF cited the second, FAR 6.302-1(a)(2)(ii), as the circumstance that applies to NextGen OPIR. FAR 6.302-1(a)(2)(ii) is as follows:

> (ii) Supplies may be deemed to be available only from the original source in the
> case of a follow-on contract for the continued development or production of a

---

[24] Note that the original Space Based Infrared System High program competed the two development contracts before engineering and manufacturing development (Space Based Infrared Systems Program Office, 2002, p. 5-4). LMS was one of the two winners and survived a down selection on November 8, 1996 (Space Based Infrared Systems Program Office, 2002, p. 3-2).

major system or highly specialized equipment, including major components thereof, when it is likely that award to any other source would result in—

(A) Substantial duplication of cost to the Government that is not expected to be recovered through competition, or

(B) Unacceptable delays in fulfilling the agency's requirements. (See 10 U.S.C. 2304 (d)(1)(B) or 41 U.S.C. 3304 (b)(2).) (FAR 6.302-1, 2019).

Using market research, the DAF determined that both (A) and (B) apply to the NextGen OPIR acquisition. After we reviewed the rest of the language in FAR 6.302, we determined that FAR 6.302-1(a)(2)(ii) and FAR 6.302-1(a)(2)(iii)—which is the same but for services instead of supplies—both appear to contain the intent that is most similar to the LTPSI approach. This specific reason applies to follow-on contracting, which is not specifically defined. Thus, it might be open to interpretation, especially within a future space architecture for which capabilities are no longer stovepiped but cut across multiple space programs covering various platforms and orbits. Regardless, this justification might be useful to understand and remember heading into future space acquisition programs that have similarities to legacy systems and could be thought of as follow-ons.

We note that although LMS received a sole-sourced contract for the GEO SVs, the DAF is still able to obtain the benefits of competition, such as innovation and lower costs, by driving LMS to regularly use open competitions for system subcomponents. Specifically, the government will require LMS to "conduct a subcontract payload competition, select up to two payload vendors, and carry them through to System [Commander], at a minimum" (AFSPC, 2018). Our overall insight is that this sole-source effort is a prudent decision because it will secure the benefits of an LTPSI approach, albeit without establishing LTPSI as a policy, and the benefits of competition at the same time.

But the J&A also reveals that the DAF is actually intending and planning for full and open competition for missile warning capabilities at the prime contractor level beginning with the fourth and fifth GEO SVs. To plan for this future competition, the DAF will sole-source the two polar SVs to Northrop Grumman, citing the industrial mobilization circumstance in FAR 6.302-3 (2019) and applying it to the need for multiple strategic satellite manufacturers. The DAF expects Northrop Grumman, which has experience in designing and manufacturing the two polar SVs, to provide a competitive proposal for the fourth and fifth GEO SVs. Thus, the inclination toward both a healthy industrial base and the status quo of full and open competition can make LTPSI difficult to pursue as a policy in the long run.

Example: The GPS IIIF

We also examined the GPS IIIF acquisition for information on when to pursue an LTPSI. The history of GPS satellites dates to the 1970s. These satellites had various models, blocks, and upgrades, such as GPS I, II, IIA, IIR, IIR-M, IIF, IIIA, and IIIF. Two contractors, Boeing and LMS, have switched back and forth at times on developing and producing these satellite series.

Boeing manufactured the IIF satellites. LMS won the contract to manufacture the IIIA satellites. In 2018, the DAF set up a two-phase competition for the IIIF satellites. Phase one was a production feasibility assessment, which Boeing, LMS, and Northrop Grumman successfully completed. Phase two was a competition for production. To the DAF's surprise, only LMS submitted a proposal; it was awarded $7.2 billion to produce 22 GPS IIIF satellites on September 14, 2018 (Divis, 2018).

Judging by industry feedback, the DAF appeared to place heavy emphasis on mature manufacturing capabilities as part of the source-selection evaluation criteria. We see how this would likely lead to a competitive proposal from LMS because it is on contract to manufacture ten GPS IIIA satellites. Therefore, a sole-source, LTPSI approach might have been more efficient. However, it is unclear what circumstances dictate whether a full and open competition or a sole-source approach is the most prudent acquisition strategy. Our initial thoughts are that this decision might depend on the following:

- whether the product or service is built to print or a replaceable element (e.g., launch)
- who has the infrastructure and expertise available
- whether the incumbent performance is unsatisfactory
- the degree to which speed is prioritized for acquisition
- strategic objectives targeting industrial base and small business mobilization.[25]

Specifically and legally, though, the decision to sole-source must meet a FAR justification. Support for this justification requires extensive market research and legal involvement to build a solid case. These practical considerations might influence the decision to pursue full and open competition as the path of least resistance. Unfortunately, this might risk yielding a nonincumbent winner that looked good on the proposal but is unable to deliver on contract.

How Often DoD Has Changed Prime Contractors

Interviews anecdotally found that prime contractors rarely changed after initial down-selection. To see whether the data uphold this belief, we conducted a quick analysis of the Selected Acquisition Reports (SARs) on 199 MDAPs since 1997 to see how often prime contractors have changed after initial award.

Table B.2 shows that out of 199, 18 mentioned multiple prime contractors in their reports, and only three were clearly cases of the prime system integrator being changed: the GPS II, the Tomahawk missile, and the Evolved Expendable Launch Vehicle. The rest appeared to use either a build-to-print or multiprime-multisystem strategy. More analysis is needed to be definitive, but this indicates that switching prime contractors is very infrequent in practice. Although the LTPSI approach can stretch beyond a defined MDAP into subsequent variants of the system that use the same architecture, these data indicate that despite regulatory barriers, it is often justifiable to keep the same system integrator in the long term. There are not absolutes (nor should there be)

---

[25] Section M of RFPs greatly determines individual sole-source selection criteria for contracts.

because contractor performance and innovation must be maintained or because some systems might be (or become) reproducible by other manufacturers. Other strategic factors, such as the retention of at least two viable contractors or production locations, also can affect a sole-source decision.

Table B.2. MDAPs with Multiple Primes (1997–2017 Selected Acquisition Reports)

Apparent Switched Prime Integrator	Multiple Primes for Other Apparent Reasons	
	Build to Print	Multiple Production Lines or Multiple Systems
• GPS II • Tomahawk missile • Evolved Expendable Launch Vehicle (EELV)	• Single-channel ground and airborne radio systems (SINCGARS) • MRAP • Manpackable radio • Thermal Weapon Sight II • Family of medium tactical vehicles (FMTV)	• Joint Capabilities Release (JCR) • T-AKR vehicle cargo ships • Sentinel/Forward Area Air Defense Command and Control (FAAD C2) • Integrated defensive electronic countermeasures (IDEC) • DDG-51 *Arleigh Burke*–class destroyer • Chemical demilitarization (2 programs) • Multifunctional Information Distribution System (MIDS) • Warfighter Simulation (WARSIM) • Littoral Combat Ship (LCS)
Totals: 3	5	10

Industrial-Base Insights from Prior RAND Research

Another potential concern for the LTPSI approach beyond reducing competition is whether it might lead to industrial-base concerns. It was beyond the scope of our project to conduct primary analysis of the defense space industrial base, but our sponsor asked us to review existing RAND research on the defense industrial base to identify relevant insights.

RAND material on the defense industrial base has focused on various issues for ship and aircraft acquisition. There are important differences between the defense industrial bases for different types of military systems, but the space acquisition community can still benefit from insights provided by these studies. We therefore review this research on the industrial base to identify relevant information that would apply to situations in which the prime contractors have a long-term tenure on space programs.

Both the ship and aircraft industrial bases have consolidated dramatically over the past several decades and are generally down to two and three major prime integration contractors, respectively. In response to this reduction, ship and aircraft acquisition programs have emphasized preserving at least two viable prime contractors through special acquisition strategies, such as split production on the same acquisition program, despite likely additional costs.

124

Although space prime contractors have also consolidated, the traditional winner-take-all acquisition environment has been sufficient in maintaining at least two prime contractors in the space industrial base. This might be because of larger numbers of both classified and unclassified satellite systems than aircraft or ship programs, but we have not analyzed the space industrial base in depth. Alternatively, winning prime contractors develop extensive design and production infrastructure for each unique satellite system (or system domain—electro-optical, infrared, signal detection, communications, etc.) and produce relatively few satellites, on the order of a few to tens. Thus, split production for space systems would come with high nonrecurring infrastructure costs and only a few end items.

These insights seem to reinforce sticking with a well-performing prime system integrator after initial competition and down-selection on the same space system bus, but there are some other considerations that are relevant. The first is the effect on costs that split production might have. Again, we note that there are differences among the industrial bases, but RAND research has generally found that split production increases costs. One study looked at three U.S. ship programs and found that the two with separate design contractors led to significant schedule and cost increases, and the program with a single design and production prime contractor was far more successful from a cost and schedule standpoint (Schank, Ip, et al., 2011). Similarly, the United Kingdom Ministry of Defence found that sole-source selection of building ships in a single shipyard is less costly than dividing production between two shipyards (Birkler, Schank, et al., 2002).

Another ship industrial base study looked at the Navy's DDG-1000 destroyer, which the Navy intended to compete throughout the design and production period. Research found, however, that reopening design rivalry after three years from start of design would result in significant costs both in time and dollars (Schank, Smith, et al., 2006). Regarding aircraft, the DoD initially considered competitive coproduction of the F-35 to drive unit price down. A study concluded that the odds are historically only 1 in 10 that the lower prices resulting from competitive coproduction of a vehicle would outweigh the extra nonrecurring costs from an additional production line (Birkler, Bower, et al., 2003). The conclusions from these RAND studies suggest that recompeting the design or production of a satellite system might not prove beneficial from a short-term cost standpoint. The final decision might depend on whether the system's architecture is proprietary or a MOSA, the learning curve for understanding the MOSA (if reopening subsequent prime-contractor awards for competition is the question), and the costs and benefits for competing components for production or upgrades. In the case of the F-35, for example, Birkler, Bower, et al. (2003) found that the historical odds for savings from competing avionics were higher (about 50-50), so component upgrades for the F-35 might be viable.

Although short-term costs are generally increased by split production or by recompeting contracts for major programs, there might be other reasons to recompete. The concern that selection of a single prime contractor for a major program would force the other to exit the market might justify recompeting a contract because the industry would cease to be competitive and long-term costs might increase (Schank, Smith, et al., 2006; Birkler, Bracken, et al., 2011).

In response to that concern, the Navy planned to split production on its DDG-1000 destroyer. Although Northrup Grumman Ship Systems was the design and proposal winner, the Navy stipulated that the other industry team participate in ship design and production to ensure continued participation in the market. A study found that as little as 33 percent of the initially planned production levels for DDG-1000 would have been sufficient to support the production base (Schank, Smith, et al., 2006). In the end, DDG-1000 quantities were reduced to three ships, so production was down-selected to General Dynamics' Bath Iron Works for the overall design, construction, integration, testing, and delivery, but split production was initially deemed viable at higher quantities (U.S. Navy, undated). Similarly, another study focused on the option of splitting final assembly and checkout sites for the F-35 and found that doing so would support the production base of the industry (Cook et al., 2002). Ultimately, split production or coproduction is a viable means of supporting the production base of the respective industry, albeit an expensive one in the short term.

An important distinction that both aforementioned reports mention is the difference between the production base and the design base. Supporting the design base might be more critical to the industry than supporting the production base is, and the reports listed several activities that could target sustaining the design base. Supporting the design base by increasing the number of programs intended for design and production seems like a simple bolstering activity but is likely infeasible because of the increased costs associated with more programs (Birkler, Bower, et al., 2003). However, a study concerned with maintaining future military aircraft design capabilities argued that increasing the number of development activities, even if they are not intended for production, would provide work to sustain the design base (Drezner et al., 1992). This insight was reiterated in two more reports, the first of which examined competition in the aircraft industry and the second looked at sustaining the United Kingdom's nuclear submarine industrial base (Birkler, Bower, et al., 2003; Schank, Riposo, et al., 2005). Alternatively, the government might assume some of the technology development and early design activities through government labs or design bureaus that are traditionally conducted by industry (Drezner et al., 1992). A final possibility would be to use spiral development to support the design base. Spiral development would mean improving the design and capabilities of the system as new technologies and threats emerge (Schank, Riposo, et al., 2005). This would keep the technical staff engaged with systems well after the first-of-kind was produced.

Ultimately, the RAND industrial base literature does not indicate whether there are sufficient reasons for split production or to recompete satellite systems after down-selection, but Arena and Birkler (2009) review some of the basic considerations. From cost, schedule, and performance standpoints, it is unlikely that recompeting would be preferable to remaining with a single incumbent prime contractor. Beyond those considerations, space programs must weigh the effects of recompeting on the space industry. Although competition after initial down-selection would likely support the production base of the industry, it would not support the design base, which prior research indicates is critical to maintaining a robust industrial base. The insights drawn from RAND's ship and aircraft industrial base literature suggest that recompeting after initial down-selection generally will incur increases in costs and cycle times without sufficient benefits. It is worth noting that the long-term prime contractor maintaining interface standards and design architecture with MOSA can enable competition and innovation at lower tiers (or primary and secondary mission systems).

Deliver-When-Mature Concept

One acquisition approach available to satellite systems is to insert capabilities as they become mature, thus allowing satellite systems to respond to threats as soon as possible. This could be accomplished either through *incremental acquisition* or through *continuous component upgrade as technology matures*. Historically, there has not been much growth in capabilities within a satellite block. Although there are advantages to this lack of growth (one of which is that fewer configurations of satellites can be easier on satellite operators), a significant disadvantage is that capabilities are not fielded soon enough, and in fact might be obsolete by the time they are incorporated into the next block. Figure B.1 demonstrates how space acquisition of satellite constellations can incorporate capability insertions to stay ahead of threats.

Figure B.1. Deliver-When-Mature Concept (Notional)

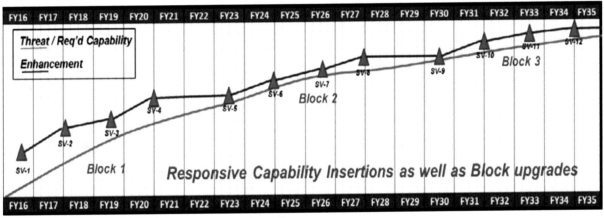

SOURCE: Unpublished DAF illustration provided to the authors.

The DAF might be able to plan up front for capability insertion by housing capability development and satellite manufacturing under the same contract. However, to maximize capability insertions, the DAF will need to rely on an approach that is able to add capabilities developed through contracts on an S&T feed. This will involve government-owned interfaces, MOSA, and other approaches. This is an exciting time for the DAF because it has set up the NextGen OPIR acquisition to meet this need. To successfully execute this deliver-when-mature concept, though, the DAF must be prepared for the challenging task of protecting its S&T feed.

127

Appendix C. Preliminary Ideas on Protecting an S&T Feed

One enabler of faster acquisition is to have a strong S&T program so that technology is mature when needed for an acquisition. As a result, the DAF asked for some initial thoughts on how such S&T feeds for acquisition might be protected.

S&T information feeds, and the budgets to support them, can be difficult to protect. What often happens is that S&T budgets are dramatically reduced once the associated program is under way and the technologies have been chosen. It can be very difficult to argue for the continuation of an S&T information feed when (1) unable to tie the investments to a specific requirement, and (2) an acquisition program is taking off and consuming a larger share of the budget every year. However, the threat landscape for space demonstrates the importance of a robust S&T budget. Therefore, we provide three initial insights that might help the DAF better defend its S&T information feeds and budgets. Further research is needed to develop these ideas.

First, the DAF could write acquisition strategies that provide a compelling business case for operating and maintaining a consistent S&T information feed throughout the life of a program or portfolio of programs. Commercial companies do this by tying their S&T efforts to a return on investment. Although profit is not a motivator, the DAF could tie S&T efforts to other benefits. For one thing, a healthy S&T feed could offer a speed advantage to the DAF. Quantitatively, the DAF might be able to determine, on average, how much time an S&T feed can save between capability required and launch. For another thing, a healthy S&T feed might allow the DAF to proactively work inside an adversary's decision loop and drive their reactions. There also might be opportunities to quantify such strategic effects and tie them to the S&T investments that enabled them. Additionally, an S&T feed that increases preparation for space warfare can serve as a space deterrent if certain parts of the S&T feed are open and transparent to adversaries (e.g., the amount of money invested). Finally, a robust S&T feed could increase the ability of the DAF to capitalize on commercial technology breakthroughs, facilitated by market analysis and foresight of the broader U.S. and allied space economy. These benefits are not all inclusive but could start a dialogue on the value of a stable space S&T feed.

Second, the DAF could characterize how its S&T information feeds fit into the overall U.S. space architecture and strategy. For example, one approach might be to have an explicit S&T budget line for each space acquisition program. Conversely, the DAF might consider constructing an aggregated S&T budget program element for a larger portfolio. The latter approach could support common S&T that cuts across various programs in the portfolio. There are advantages and disadvantages to each approach. One disadvantage to the latter approach is that an aggregated, large S&T budget could be a major target for budget cuts (i.e., large budget items for broader purposes are harder to defend than smaller budget items with closer ties to specific space capabilities, acquisition strategies, and programs).

Third, and finally, the DAF could explore ways to empower key leaders, both within the acquisition decision chain of command (e.g., the PEO or the Director of the NRO) and external to it (e.g., the air staff requirements community), to review these S&T justifications in detail and offer stable support to S&T and PMs. Because these key leaders have coverage across a variety of programs, they should have direct input into future S&T investments for both individual programs and the entire portfolio. Having a strong and consistent voice for an S&T feed that can echo across all three acquisition legs (defense acquisition system, budget, and requirements) can be critical to defending the S&T budget.

Appendix D. Using Decision Charts to Select Approaches

Basic decision charts can help to partition the option space into groups of potential approaches, but we found that the factors involved tend to be too numerous to isolate the candidate approaches into small numbers. Figure D.1 illustrates a potential decision process based on a flowchart selection method. Here, answering three to four questions about a particular acquisition sends the user to a set of potential candidate approaches (either organizational models with their associated sets of approaches or a specific approach). The challenge is that we still have nuanced differences among the resulting options, and critical thinking is required. Also, the list of specific approaches does not span the full list in Box 2.1. As a result, we developed an expanded selection method described in the main body of the report and illustrated in Figure 3.1.

Figure D.1 shows an initial approach used to explore how various basic factors could affect the strategies used to acquire capabilities rapidly. At the same time, various organizations were placed at the endpoints to see if solution sets seemed strongly clustered. Some patterns did emerge. If a capability is needed in a very short time frame, the options are limited: (1) COTS acquisition of an existing, mature, system (Iridium, MRAP) or (2) repurposing (reprogramming) an existing system to perform another mission. Programs for rapid development of large, expensive systems tended to look like Skunk Works or DAF RCO (which was modeled on Skunk Works).

Figure D.1. Potential Flow Diagram for Selecting Acceleration Approaches

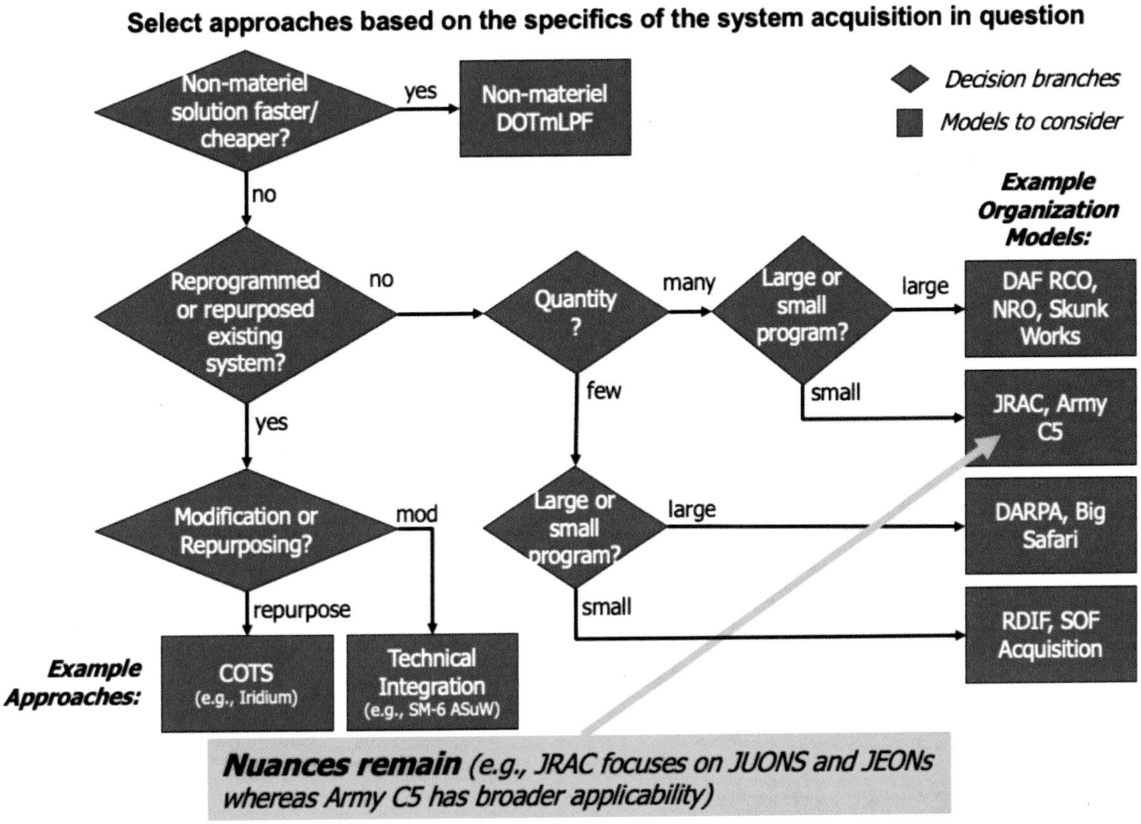

Select approaches based on the specifics of the system acquisition in question

NOTE: DOTmLPF = doctrine, organization, training, materiel, leadership and education, personnel, and facilities.

However, the model is not granular enough to separate out how organizations that fulfill somewhat similar acquisition functions accomplish it in radically different ways or have radically different portfolios. For instance, DARPA and Big Safari both produce small numbers of sensitive equipment, but Big Safari is focused on integration of existing COTS technology, and DARPA looks for revolutionary technology breakthroughs. JRAC and Army C5 both focus on rapid prototyping but employ radically different strategies and administrative structures (JRAC is part of JCIDS, C5 avoids it altogether).

Thus, a finer parsing of individual elements was necessary to categorize approaches and to build a model by which a user could decide what approaches would be most appropriate to their specific program. This led to the development of the framework discussed in Chapters 2–3.

References

88th Air Base Wing Public Affairs, "Collaboration Delivers Capability to the Warfighter," Wright-Patterson Air Force Base webpage, February 16, 2018. As of November 13, 2018: https://www.wpafb.af.mil/News/Article-Display/Article/1443456/collaboration-delivers-capability-to-the-warfighter/

Advisory Panel on Streamlining and Codifying Acquisition Regulations, *Report of the Advisory Panel on Streamlining and Codifying Acquisition Regulations,* Vol. 2 of 3, Arlington, Va., June 2018. As of November 26, 2018: https://section809panel.org/wp-content/uploads/2018/07/Sec809Panel_Vol2-Report_June2018.pdf

Aerospace Corporation, "Hive Satellites Redefine Disaggregation," press release, February 8, 2018. As of December 11, 2018: https://aerospace.org/story/hive-satellites-redefine-disaggregation

AFI—*See* Air Force Instruction.

AFSPC—*See* Air Force Space Command.

Air Force Instruction 10-601, *Operational Capability Requirements Development*, Washington, D.C.: U.S. Air Force, November 6, 2013. As of December 9, 2019: https://static.e-publishing.af.mil/production/1/af_a3_5/publication/afi10-601/afi10-601.pdf

Air Force Instruction 63-101/20-101, *Integrated Life Cycle Management*, Washington, D.C.: U.S. Air Force, May 9, 2017. As of November 28, 2018: https://static.e-publishing.af.mil/production/1/saf_aq/publication/afi63-101_20-101/afi63-101_20-101.pdf

Air Force Space Command, *Resiliency and Disaggregated Space Architectures*, white paper, April 14, 2016. As of October 24, 2018: https://www.afspc.af.mil/Portals/3/documents/AFD-130821-034.pdf?ver=2016

———, "Next Generation Overhead Persistent Infrared (Next Gen OPIR) Geosynchronous Earth Orbit (GEO) Space Vehicles 1–3" (redacted), Los Angeles Air Force Base, El Segundo, Calif.: Department of the Air Force, August 27, 2018.

Ambrose, Rick, "Four Predictions for the Future of Space," *LinkedIn,* March 29, 2018a. As of October 25, 2018: https://www.linkedin.com/pulse/four-predictions-future-space-rick-ambrose?trk=portfolio_article-card_title

———, "Why 'Yes, and . . .' Will Mark the Future of Space," *LinkedIn*, April 20, 2018b. As of October 25, 2018:
https://www.linkedin.com/pulse/why-yes-mark-future-space-rick-ambrose?trk=portfolio_article-card_title

———, "How Satellites Will Turn into Smartphones," *LinkedIn*, June 1, 2018c. As of October 25, 2018:
https://www.linkedin.com/pulse/how-satellites-turn-smartphones-rick-ambrose?trk=portfolio_article-card_title

———, "How We're Striving to Build Satellites Twice as Fast and at Half the Cost," LinkedIn, June 26, 2018d. As of October 25, 2018:
https://www.linkedin.com/pulse/how-were-striving-build-satellites-twice-fast-half-cost-rick-ambrose

Arena, Mark V., and John Birkler, *Determining When Competition Is a Reasonable Strategy for the Production Phase of Defense Acquisition*, Santa Monica, Calif.: RAND Corporation, OP-263-OSD, 2009. As of October 11, 2018:
https://www.rand.org/pubs/occasional_papers/OP263.html

Assistant Secretary of the Air Force for Acquisition, Technology, and Logistics, "Air Force Guidance Memorandum for Rapid Acquisition Activities," AFGM2018-63-146-01, Washington, D.C.: U.S. Department of Defense, Department of the Air Force, June 13, 2018.

Bender, Bryan, and Jacqueline Klimas, "Trump's Space Force Struggling to Launch," *Politico,* September 17, 2018. As of December 6, 2018:
https://www.politico.com/story/2018/09/17/space-force-military-air-force-825757

Berkowitz, Bruce, "Lessons Learned, Organizational Culture, and the Future of the National Reconnaissance Office," *National Reconnaissance—Journal of the Discipline and Practice*, Chantilly, Va.: National Reconnaissance Office, January 2015.

Birkler, John, Anthony G. Bower, Jeffrey A. Drezner, Gordon T. Lee, Mark A. Lorell, Giles K. Smith, Fred Timson, William P. G. Trimble, and Obaid Younossi, *Competition and Innovation in the U.S. Fixed-Wing Military Aircraft Industry*, Santa Monica, Calif.: RAND Corporation, MR-1656-OSD, 2003. As of July 27, 2018:
https://www.rand.org/pubs/monograph_reports/MR1656.html

Birkler, John, Paul Bracken, Gordon T. Lee, Mark A. Lorell, Soumen Saha, and Shane Tierney, *Keeping a Competitive U.S. Military Aircraft Industry Aloft: Findings from an Analysis of the Industrial Base*, Santa Monica, Calif.: RAND Corporation, MG-1133-OSD, 2011. As of May 25, 2018:
https://www.rand.org/pubs/monographs/MG1133.html

Birkler, John, John F. Schank, Mark V. Arena, Giles K. Smith, and Gordon T. Lee, *The Royal Navy's New-Generation Type 45 Destroyer: Acquisition Options and Implications*, Santa Monica, Calif.: RAND Corporation, MR-1486-MOD, 2002. As of May 25, 2018: https://www.rand.org/pubs/monograph_reports/MR1486.html

Boyle, Alan, "Transcript: 'Chief Slowdown Officer' Jeff Bezos Shares Amazon Management Tips," *Geekwire*, September 19, 2018. As of October 25, 2018: https://www.geekwire.com/2018/full-transcript-chief-slowdown-officer-jeff-bezos-shares-amazon-management-wisdom/

Brooks, Frederick P., Jr., *The Mythical Man-Month: Essays on Software Engineering*, Reading, Mass.: Addison-Wesley, 1975.

———, *The Mythical Man-Month: Essays on Software Engineering*, anniversary edition, Boston, Mass.: Addison Wesley, 1995.

Camm, Frank, Brian K. Dougherty, and Thomas C. Whitmore, *Improving Acquisition to Support the Space Enterprise Vision: Supplemental Appendixes on Acquisition Concepts*, Santa Monica, Calif.: RAND Corporation, RR-2626/1-AF, forthcoming.

Carter, Ash, "Drell Lecture: 'Rewiring the Pentagon: Charting a New Path on Innovation and Cybersecurity' (Stanford University)," U.S. Department of Defense webpage, April 23, 2015. As of November 13, 2018: https://www.defense.gov/Newsroom/Speeches/Speech/Article/606666/drell-lecture-rewiring-the-pentagon-charting-a-new-path-on-innovation-and-cyber/

Chairman of the Joint Chiefs of Staff Instruction 5123.01H, *Charter of the Joint Requirements Oversight Council (JROC) and Implementation of the Joint Capabilities Integration and Development System (JCIDS)*, Washington, D.C.: Joint Staff, August 31, 2018. As of May 2, 2019: https://www.jcs.mil/LinkClick.aspx?fileticket=1izOP9Dwxqs%3d&tabid=19767&portalid=36&mid=46626

Chen, Angela, "The Commercialization of Space," *JSTOR Daily,* April 28, 2016. As of December 6, 2018: https://daily.jstor.org/commercialization-of-space/

Cheung, Tai Ming, "Innovation in China's Defense Technology Base: Foreign Technology and Military Capabilities," *Journal of Strategic Studies*, Vol. 39, Nos. 5–6, September 11, 2018, pp. 728–761.

CJCSI—*See* Chairman of the Joint Chiefs of Staff Instruction.

Clark, Colin, "F-35 Production Move Was 'Acquisition Malpractice': Top DoD Buyer," *Breaking Defense*, February 6, 2012. As of June 7, 2019:

https://breakingdefense.com/2012/02/f-35-production-was-acquisition-malpractice-top-dod-weapons-b/

Clark, James W., Jr., *Acquisition Streamlining: A Viable Method for Accelerated Procurement of the Advanced Amphibious Assault Vehicle*, master's thesis, Monterey, Calif.: Naval Postgraduate School, December 1993. As of June 6, 2019:
https://apps.dtic.mil/dtic/tr/fulltext/u2/a276429.pdf

Code of Federal Regulations, Title 32, National Defense, Subtitle A, Department of Defense, Chapter 1, Office of the Secretary of Defense, Subchapter C, DoD Grant and Agreement Regulations, Part 37, Technology Investment Agreements, August 7, 2003.

Consortium Management Group, "C5 Membership List," webpage, undated. As of November 13, 2018:
https://cmgcorp.org/c5/member-list/

Cook, Cynthia R., Mark V. Arena, John Graser, John A. Ausink, Lloyd Dixon, Timothy Liston, Sheila E. Murray, Susan A. Resetar, Chad Shirley, Jerry M. Sollinger, and Obaid Younossi, *Final Assembly and Checkout Alternatives for the Joint Strike Fighter*, Santa Monica, Calif.: RAND Corporation, MR-1559-OSD, 2002. As of July 20, 2018:
https://www.rand.org/pubs/monograph_reports/MR1559.html

Dana, W. H., *The X-15 Lessons Learned*, NASA Dryden Research Facility, technical report, 1993.

DARPA—*See* Defense Advanced Research Projects Agency.

DASD(SE)—*See* Deputy Assistant Secretary of Defense for Systems Engineering.

DAU—*See* Defense Acquisition University.

Davis, Lorrie A., and Lucien Filip, *How Long Does It Take to Develop and Launch Government Satellite Systems?* Los Angeles, Calif.: Aerospace Corporation, ATR-2015-00535, March 12, 2015.

Defense Acquisition University, "Adaptive Acquisition Framework," webpage, undated-a. As of June 26, 2019:
https://aaf.dau.edu/aaf

———, "Joint Rapid Acquisition Cell," *Defense Acquisition Glossary*, undated-b. As of December 11, 2018:
https://www.dau.edu/glossary/Pages/Glossary.aspx#!both|J|27781

———, *Manager's Guide to Technology Transition in an Evolutionary Acquisition Environment*, Version 2.0, Fort Belvoir, Va.: Defense Acquisition University Press, June 2005. As of June 19, 2019:
https://apps.dtic.mil/dtic/tr/fulltext/u2/a484102.pdf

———, *Defense Acquisition Structures and Capabilities Review Addendum, Pursuant to Section 814, National Defense Authorization Act, Fiscal Year 2006*, June 2007. As of November 28, 2018:
https://apps.dtic.mil/dtic/tr/fulltext/u2/a524299.pdf

Defense Advanced Research Projects Agency, "Contract Management," webpage, Washington, D.C.: U.S. Department of Defense, undated-a. As of November 20, 2018:
https://www.darpa.mil/work-with-us/contract-management

———, homepage, undated-b. As of December 11, 2018:
https://www.darpa.mil

———, "Other Transactions for Prototypes Fact Sheet," Washington, D.C.: U.S. Department of Defense, October 19, 2018. As of December 10, 2019:
https://www.darpa.mil/attachments/SBIR-OT-Fact-Sheet-19-Oct-18.pdf

Defense Advanced Research Projects Agency Small Business Programs Office, *Transition & Commercialization Strategy Development Guide,* Arlington, Va.: U.S. Department of Defense, January 3, 2018. As of June 19, 2019:
https://www.darpa.mil/attachments/Transition-and-Commercialization-Guide.pdf

Defense Federal Acquisition Regulation Supplement, Acquisition.gov, October 1, 2019. As of January 2, 2020:
https://www.acquisition.gov/dfars

Defense Innovation Unit, homepage, undated. As of November 13, 2018:
https://www.diu.mil/

———, *Defense Innovation Unit (DIU) Annual Report 2018*, Silicon Valley, Calif.: U.S. Department of Defense, 2018. As of December 11, 2019:
https://www.diu.mil/download/datasets/2278/DIU%202018_Annual%20Report_FINAL.pdf

Defense Innovation Unit Experimental, *Defense Innovation Unit Experimental (DIUx) Annual Report 2017*, Silicon Valley, Calif.: U.S. Department of Defense, 2017. As of June 19, 2019:
https://www.diu.mil/download/datasets/1774/DIUx%20Annual%20Report%202017.pdf

Defense Intelligence Agency, *Russia Military Power: Building a Military to Support Great Power Aspirations*, Washington, D.C., June 23, 2017. As of December 6, 2018:
http://www.dia.mil/Portals/27/Documents/News/Military%20Power%20Publications/Russia%20Military%20Power%20Report%202017.pdf

Defense Logistics Agency, "Electronic Catalog (ECAT)," webpage, undated. As of December 11, 2018:
http://www.dla.mil/TroopSupport/Medical/ECAT.aspx

Defense Procurement and Acquisition Policy, *Manager's Guide to Technology Transition in an Evolutionary Acquisition Environment,* Version 1.0, Washington, D.C.: Office of the Under

Secretary of Defense (Acquisition, Technology, and Logistics), January 31, 2003. As of June 19, 2019:
https://www.acq.osd.mil/dpap/Docs/AQ201S1v10Complete.pdf

———, "Guidance on Using Incentive and Other Contract Types," memorandum, Washington, D.C., April 1, 2016. As of November 26, 2018:
https://www.acq.osd.mil/dpap/policy/policyvault/USA001270-16-DPAP.pdf

Defense Systems Management College, *Schedule Guide for Program Managers*, Ft. Belvoir, Va.: Defense Systems Management College Press, October 2001.

Deputy Assistant Secretary of Defense for Systems Engineering, *Department of Defense Risk, Issue, and Opportunity Management Guide for Defense Acquisition Programs*, Washington, D.C.: U.S. Department of Defense, January 2017. As of November 26, 2018:
https://www.acq.osd.mil/se/docs/2017-rio.pdf

DFARS—*See* Defense Federal Acquisition Regulation Supplement.

Dillon, Robin L., and Peter Madsen, "The Legacy of Faster-Better-Cheaper: Too Much Risk or Over-Reaction to Perceived Failure?" *2014 IEEE Aerospace Conference*, Big Sky, Mont., IEEE, March 1–8, 2014. As of November 12, 2018:
https://ieeexplore.ieee.org/stamp/stamp.jsp?arnumber=6836168

———, "Faster-Better-Cheaper Projects: Too Much Risk or Overreaction to Perceived Failure?" *IEEE Transactions on Engineering Management*, Vol. 62, No. 2, May 2015, pp. 141–149. As of November 12, 2018:
https://ieeexplore.ieee.org/document/7055338

DIU—*See* Defense Innovation Unit.

DIUx—*See* Defense Innovation Unit Experimental.

Divis, Dee Ann, "Lockheed Awarded $7.2 billion GPS IIIF Contract," *Inside GNSS*, September 15, 2018. As of October 26, 2018:
http://insidegnss.com/lockheed-awarded-7-2-billion-gps-iiif-contract/

DoDD—*See* DoD Directive.

DoD Directive 5000.01, *The Defense Acquisition System*, May 12, 2003, Incorporating Change 2, Washington, D.C.: U.S. Department of Defense, August 31, 2018. As of December 9, 2019:
https://www.esd.whs.mil/Portals/54/Documents/DD/issuances/dodd/500001p.pdf?ver=2018-09-28-073203-530

DoD Directive 5000.71, *Rapid Fulfillment of Combatant Commander Urgent Operational Needs*, August 24, 2012, Incorporating Change 1, Washington, D.C.: U.S. Department of Defense, August 31, 2018.

DoD Financial Management Regulation 7000.14-R, Vol. 2b, Budget Formulation and Presentation; Chapter 5, Research, Development, Test, and Evaluation Appropriations; Washington, D.C., U.S. Department of Defense, November 2017. As of January 22, 2020: https://comptroller.defense.gov/Portals/45/documents/fmr/Volume_02b.pdf

DoD Instruction 5000.02, *Operation of the Defense Acquisition System*, January 7, 2015, Incorporating Change 4, Washington, D.C.: U.S. Department of Defense, August 31, 2018.

DoDI—*See* DoD Instruction.

DoD Manual 5000.78, *Rapid Acquisition Authority (RAA)*, Washington, D.C.: U.S. Department of Defense, March 20, 2019. As of December 10, 2019: https://www.esd.whs.mil/Portals/54/Documents/DD/issuances/dodm/500078m.pdf?ver=2019-03-21-080447-490

DPAP—*See* Defense Procurement and Acquisition Policy.

Drezner, Jeffrey A., Giles K. Smith, Lucille E. Horgan, J. Curt Rogers, and Rachel Schmidt, *Maintaining Future Military Aircraft Design Capability*, Santa Monica, Calif.: RAND Corporation, R-4199-AF, 1992. As of May 25, 2018: https://www.rand.org/pubs/reports/R4199.html

DSMC—*See* Defense Systems Management College.

Dugan, Regina E., and Kaigham J. Gabriel, "'Special Forces' Innovation: How DARPA Attacks Problems," *Harvard Business Review*, October 2013. As of November 13, 2018: https://hbr.org/2013/10/special-forces-innovation-how-darpa-attacks-problems

Erwin, Sandra, "SMC 2.0: Air Force Begins Major Reorganization of Acquisition Offices," *Space News*, April 17, 2018a. As of December 6, 2018: https://spacenews.com/smc-2-0-air-force-begins-major-reorganization-of-acquisition-offices/

———, "First Order of Business for Air Force Space Innovation Office: Decide What It Wants to Build," *Space News*, June 25, 2018b. As of November 13, 2018: https://spacenews.com/first-order-of-business-for-air-force-space-innovation-office-decide-what-it-wants-to-build/

———, "Clearer Picture Emerging of the Future of 'Rapid Space,'" *Space News*, July 17, 2018c. As of December 10, 2018: https://spacenews.com/clearer-picture-emerging-of-the-future-of-rapid-space/

FAR—*See* Federal Acquisition Regulation.

Farnsworth, Wesley, "Wright-Patterson Facility Saves DoD Time, Millions," Air Force Materiel Command website, December 7, 2015. As of November 20, 2018: https://www.afmc.af.mil/News/Article-Display/Article/803763/wright-patterson-facility-saves-dod-time-millions/

Federal Acquisition Regulation Part 5, Publicizing Contract Actions, Acquisitions.gov, October 10, 2019. As of October 31, 2019:
https://www.acquisition.gov/content/part-5-publicizing-contract-actions

Federal Acquisition Regulation Part 6, Competition Requirements, Acquisitions.gov, October 10, 2019. As of October 31, 2019:
https://www.acquisition.gov/content/part-6-competition-requirements

Federal Acquisition Regulation, Part 30, Cost Accounting Standards, Acquisitions.gov, October 10, 2019. As of October 31, 2019:
https://www.acquisition.gov/content/part-30-cost-accounting-standards-administration

Federal Acquisition Regulation 6.302, Circumstances Permitting Other Than Full and Open Competition, Acquisitions.gov, October 10, 2019. As of October 31, 2019:
https://www.acquisition.gov/content/6302-circumstances-permitting-other-full-and-open-competition

Federal Acquisition Regulation 6.302-1, Only One Responsible Source and No Other Supplies or Services Will Satisfy Agency Requirements, Acquisitions.gov, October 10, 2019. As of October 31, 2019:
https://www.acquisition.gov/content/6302-1-only-one-responsible-source-and-no-other-supplies-or-services-will-satisfy-agency-requirements

Federal Acquisition Regulation 6.302-3, Industrial Mobilization; Engineering, Developmental, or Research Capability; or Expert Services, Acquisitions.gov, October 10, 2019. As of October 31, 2019:
https://www.acquisition.gov/content/6302-1-only-one-responsible-source-and-no-other-supplies-or-services-will-satisfy-agency-requirements

Federal Acquisition Regulation Subpart 16.4, Incentive Contracts, Acquisitions.gov, October 10, 2019. As of October 31, 2019:
https://www.acquisition.gov/content/subpart-164-incentive-contracts

Federal Acquisition Regulation Subpart 16.5, Indefinite Delivery Contracts, Acquisitions.gov, October 10, 2019. As of October 31, 2019:
https://www.acquisition.gov/content/subpart-165-indefinite-delivery-contracts

Federal Acquisition Regulation 16.501-1, Definitions, Acquisitions.gov, October 10, 2019. As of October 31, 2019:
https://www.acquisition.gov/content/16501-1-definitions

Federal Acquisition Regulation 16.501-2, General, Acquisitions.gov, October 10, 2019. As of October 31, 2019:
https://www.acquisition.gov/content/16501-2-general

Federal Acquisition Regulation 16.603, Letter Contracts, Acquisitions.gov, October 10, 2019. As of October 31, 2019:
https://www.acquisition.gov/content/16603-letter-contracts

Felt, Eric J., *Cost Considerations of Transition Toward a Disaggregated Satellite Architecture*, Maxwell Air Force Base, Ala.: Air War College, Technical Report, February 14, 2013.

Furin, Timothy A., "Bid Protests: Are Other Transaction Agreements (OTAs) Really Bulletproof?" *Federal Construction Contracting Blog*, Philadelphia, Pa: Cohen Seglias Pallas Greenhall & Furman PC, July 10, 2018. As of June 3, 2019:
https://federalconstruction.phslegal.com/2018/07/articles/bid-protests/bid-protests-are-other-transaction-agreements-otas-really-bulletproof/

GAO—*See* U.S. Government Accountability Office.

Gilb, Tom, *Software Metrics*, New York: Little, Brown, and Co., 1977.

Grimes, Bill, *The History of Big Safari*, Bloomington, Ind.: Archway Publishing, 2014.

Gruss, Mike, "Disaggregation Giving Way to Broader Space Protection Strategy," *SpaceNews*, April 26, 2015. As of October 24, 2018:
https://spacenews.com/disaggregation-giving-way-to-broader-space-protection-strategy/

GSA—*See* U.S. General Services Administration.

Hagan, Gary, "Transition of Technology into the DoD Acquisition Process," briefing slides, DARPA SBIR Phase I Training Workshop, Ft. Belvoir, Va.: Defense Acquisition University, May 4, 2011. As of June 19, 2019:
https://www.darpa.mil/attachments/(6T5)%20Global%20Nav%20-%20Work%20With%20Us%20-%20For%20Small%20Business%20-%20Resource%20(Approved).pdf

———, *DAU Glossary of Defense Acquisition Acronyms & Terms,* 16th ed., Fort Belvoir, Va.: Defense Acquisition University Press, September 2015. As of December 9, 2019:
https://www.dau.edu/glossary/Documents/Glossary_16th%20_ed.pdf

Harrington, John, "Changing the Face of War; Saving Lives—The Legacy of Bill Grimes," Wright-Patterson Air Force Base website, September 24, 2018. As of December 11, 2018:
https://www.wpafb.af.mil/News/Article-Display/Article/1643188/changing-the-face-of-war-saving-lives-the-legacy-of-bill-grimes/

Insinna, Valerie, "Pentagon Presents Recommendations on Space Force to Trump," *Defense News*, October 23, 2018. As of December 6, 2018:
https://www.defensenews.com/space/2018/10/23/pentagon-presents-recommendations-on-space-force-to-trump/

Inspector General of the National Aeronautics and Space Administration, *Faster, Better, Cheaper: Policy, Strategic Planning, and Human Resource Alignment*, Washington, D.C.: National Aeronautics and Space Administration, Audit Report IG-01-009, March 13, 2001.

Kendall, Frank, "The Original Better Buying Power—David Packard Acquisition Rules 1971," *Defense AT&L*, May–June 2013. As of December 9, 2019:
https://www.dau.edu/library/defense-atl/DATLFiles/May-Jun2013/Kendall.pdf

———, *Getting Defense Acquisition Right*, Fort Belvoir, Va.: Defense Acquisition University Press, 2017.

Kim, Yool, Guy Weichenberg, Frank Camm, Brian K. Dougherty, Thomas C. Whitmore, Nicholas Martin, and Badreddine Ahtchi, *Improving Acquisition to Support the Space Enterprise Vision*, Santa Monica, Calif.: RAND Corporation, RR-2626-AF, forthcoming.

Klimas, Jacqueline, "Speeding Up Space Acquisition 'Number One Issue,' White House Adviser Says," *Politico*, June 22, 2018. As of December 6, 2018:
https://www.politico.com/story/2018/06/22/space-eric-stallmer-commercial-spaceflight-foundation-664284

Kossiakoff, Alexander, William N. Sweet, Samuel J. Seymour, and Steven M. Biemer, *Systems Engineering Principles and Practice*, Hoboken, N.J.: Wiley, 2011.

LaGrone, Sam, "Navy Sinks Former Frigate USS Reuben James in Test of New Supersonic Anti-Surface Missile," *USNI News*, Annapolis, Md.: U.S. Naval Institute, March 8, 2016. As of December 12, 2018:
https://news.usni.org/2016/03/07/navy-sinks-former-frigate-uss-reuben-james-in-test-of-new-supersonic-anti-surface-missile

Lapham, Mary Ann, Suzanne Miller, Lorraine Adams, Nanette Brown, Bart Hackemack, Charles (Bud) Hammons, Linda Levine, and Alfred Schenker, *Agile Methods: Selected DoD Management and Acquisition Concerns*, Pittsburgh, Pa: Carnegie Mellon University, October 2011. As of June 6, 2019:
https://resources.sei.cmu.edu/asset_files/TechnicalNote/2011_004_001_15335.pdf

Lapham, Mary Ann, Ray Williams, Charles (Bud) Hammons, Daniel Burton, and Alfred Schenker, *Considerations for Using Agile in DoD Acquisition*, Pittsburgh, Pa: Carnegie Mellon University, April 2010. As of June 6, 2019:
https://resources.sei.cmu.edu/asset_files/TechnicalNote/2010_004_001_15155.pdf

Larman, Craig, and Victor R. Basili, "Iterative and Incremental Development: A Brief History," *Computer*, Vol. 36, No. 6, June 2003, pp. 47–56.

Lasky, Adam K., "Choosing the Best Forum for Filing Your Bid Protest—GAO vs. Court of Federal Claims," *The Procurement Playbook*, blog post, February 2, 2017.

Levinson, Robert, "Pentagon Eyes 'Tiny' Rockets for Small Reconnaissance Satellites," Bloomberg, April 4, 2018.

Lockheed Martin, "Kelly Johnson's 14 Rules and Practices," webpage, undated. As of November 13, 2018:
https://www.lockheedmartin.com/en-us/who-we-are/business-areas/aeronautics/skunkworks/kelly-14-rules.html

Lorell, Mark A., Julia F. Lowell, and Obaid Younossi, *Evolutionary Acquisition: Implementation Challenges for Defense Space Program*, Santa Monica, Calif.: RAND Corporation, MG-431-AF, 2006. As of June 6, 2019:
https://www.rand.org/pubs/monographs/MG431.html

Loudin, Kathlyn Hopkins, "Lead Systems Integrators: A Post-Acquisition Reform Retrospective," *Acquisition Research Journal*, Ft. Belvoir, Va: Defense Acquisition University, January 2010.

Makufka, David, "Kennedy Space Center *Swamp Works*—Developing New Tools for Deep Space Exploration," Technology Transfer Office, Kennedy Space Center, Fla.: National Aeronautics and Space Administration, undated. As of June 4, 2019:
https://technology-ksc.ndc.nasa.gov/featurestory/swampworks

Mantel, Samuel J., Jr., Jack R. Meredith, Scott M. Shafer, and Margaret M. Sutton, *Project Management in Practice*, 4th ed., Hoboken, N.J.: John Wiley & Sons, 2011.

Markuson, Nancy, and Joleen Flasher, "Adapting Agile Processes for Military Acquisition Programs," briefing, McLean, Va.: MITRE Corp., April 1, 2014.

McKernan, Megan, Jeffrey A. Drezner, and Jerry M. Sollinger, *Tailoring the Acquisition Process in the U.S. Department of Defense*, Santa Monica, Calif.: RAND Corporation, RR-966-OSD, 2015. As of June 6, 2019:
https://www.rand.org/pubs/research_reports/RR966.html

McNutt, Ross, *Reducing DoD Product Development Time: The Role of the Schedule Development Process*, doctoral dissertation, Cambridge, Mass.: Massachusetts Institute of Technology, December 1998.

Mehta, Aaron, "House Committee Explores Ending Strategic Capabilities Office," *Defense News*, April 25, 2018. As of November 27, 2018:
https://www.defensenews.com/congress/2018/04/25/house-committee-explores-ending-strategic-capabilities-office/

Miller, Jay, *Lockheed Martin's Skunk Works*, Leicester, England: Midland Publishing Ltd., 1995.

Miller, Suzanne, Dan Ward, Mary Ann Lapham, Ray Williams, Charles (Bud) Hammons, Daniel Burton, and Alfred Schenker, *Update 2016: Considerations for Using Agile in DoD Acquisition*, Pittsburgh, Pa: Carnegie Mellon University, December 2016. As of June 6,

2019:
https://resources.sei.cmu.edu/asset_files/TechnicalNote/2016_004_001_484651.pdf

MITRE Corp., *Systems Engineering Guide*, Bedford, Mass., 2014. As of June 3, 2019:
http://www.mitre.org/sites/default/files/publications/se-guide-book-interactive.pdf

Modigliani, Pete, and Su Chang, *Defense Agile Acquisition Guide: Tailoring DoD IT Acquisition Program Structures and Processes to Rapidly Deliver Capabilities*, McLean, Va.: MITRE Corp., March 2014. As of October 1, 2018:
http://www.mitre.org/publications/technical-papers/defense-agile-acquisition-guide-tailoring-dod-it-acquisition-program

Moeller, William G., *Accelerating the Decision Process in Major System Acquisition*, Washington, D.C.: Logistics Management Institute, September 1979. As of June 6, 2019:
https://apps.dtic.mil/dtic/tr/fulltext/u2/a078326.pdf

National Defense Industrial Association, *Planning & Scheduling Excellence Guide (PASEG), Published Release v2.0*, Arlington, Va.: Procurement Division—Program Management Systems Committee, June 22, 2012.

———, *Planning & Scheduling Excellence Guide (PASEG), Version 3.0*, Arlington, Va.: Integrated Program Management Division, March 9, 2016. As of June 7, 2019:
http://www.ndia.org/-/media/sites/ndia/meetings-and-events/divisions/ipmd/links-and-reference/planning-and-scheduling-excellence-guide-paseg.ashx?la=en

National Reconnaissance Office, homepage, undated-a. As of December 11, 2018:
www.nro.gov/

———, "NRO Director's Innovation Initiative," webpage, undated-b. As of June 5, 2019:
http://www.nro.gov/Business-Innovation-Opportunities/

———, "National Reconnaissance Office Broad Agency Announcement for the FY2020 Director's Innovation Initiative Program," Solicitation Number: NRO000-19-R-0222, fbo.gov, Washington, D.C.: U.S. General Services Administration, May 17, 2019.

National Research Council, *Evaluation of U.S. Air Force Preacquisition Technology Development*, Washington, D.C.: National Academies Press, 2011.

NDIA—*See* National Defense Industrial Association.

Nemfakos, Charles, Irv Blickstein, Aine Seitz McCarthy, and Jerry M. Sollinger, *The Perfect Storm: The Goldwater-Nichols Act and Its Effect on Navy Acquisition*, Santa Monica, Calif.: RAND Corporation, OP-308-NAVY, 2010. As of November 28, 2018:
https://www.rand.org/pubs/occasional_papers/OP308.html

Northern, Carlton, Kathleen Mayfield, Robert Benito, and Michelle Casagni, *Handbook for Implementing Agile in Department of Defense Information Technology Acquisition*, McLean, Va.: MITRE Corp., Technical Report MTR100489, December 15, 2010.

NRO—*See* National Reconnaissance Office.

Oar, Amber R., John P. Fitzsimmons, James P. Guthrie, Raymond A. Hoffman, Andrew J. Metzger, Carl J. Nelson, Taylor J. Olson, and Matthew J. Postupack, "DoD Acquisitions Reform: Embracing and Implementing Agile," *Air and Space Power Journal*, Vol. 29, No. 6. December 2015. As of June 6, 2019:
https://www.airuniversity.af.edu/Portals/10/ASPJ/journals/Volume-29_Issue-6/DoD_Acquisitions_Reform.pdf

Oberg, James, "NASA Study: Faster, Better, Cheaper Fails," UPI, March 15, 2000. As of December 15, 2019:
https://www.upi.com/Archives/2000/03/15/NASA-study-faster-better-cheaper-fails/3418953096400/

Office of the Secretary of Defense, *Annual Report to Congress: Military and Security Developments Involving the People's Republic of China 2018*, Washington, D.C., May 16, 2018a. As of December 6, 2018:
https://media.defense.gov/2018/Aug/16/2001955282/-1/-1/1/2018-CHINA-MILITARY-POWER-REPORT.PDF

———, *Summary of the 2018 National Defense Strategy of the United States: Sharpening the American Military's Competitive Edge*, Washington, D.C.: U.S. Department of Defense, 2018b. As of June 19, 2018:
https://dod.defense.gov/Portals/1/Documents/pubs/2018-National-Defense-Strategy-Summary.pdf

OSD—*See* Office of the Secretary of Defense.

Public Law 107-314, Bob Stump National Defense Authorization Act for Fiscal Year 2003, December 2, 2002.

Public Law 114-92, National Defense Authorization Act for Fiscal Year 2016, November 25, 2015. As of December 11, 2018:
https://www.congress.gov/114/plaws/publ92/PLAW-114publ92.pdf

Public Law 114-328, National Defense Authorization Act for Fiscal Year 2017, December 23, 2016. As of October 17, 2019:
https://uscode.house.gov/statviewer.htm?volume=130&page=2256

Public Law 115-91, National Defense Authorization Act for Fiscal Year 2018, December 12, 2017. As of December 11, 2018:
https://www.congress.gov/115/plaws/publ91/PLAW-115publ91.pdf

Public Law 115-232, John S. McCain National Defense Authorization Act for Fiscal Year 2019, August 13, 2018. As of October 31, 2019:
https://www.congress.gov/115/plaws/publ232/PLAW-115publ232.pdf

Public Law 116-92, National Defense Authorization Act for Fiscal Year 2020, December 20, 2019.

Randell, B., and F. W. Zurcher, "Iterative Multi-Level Modeling: A Methodology for Computer System Design," *Proceedings of IFIP Congress*, Edinburgh, 1968, pp. 867–871.

Reagan, Rex B., and David F. Rico, "Lean and Agile Acquisition and Systems Engineering: A Paradigm Whose Time Has Come," *Defense AT&L*, November–December 2010, pp. 49–52.

Reed, Scott, and Kathryn Ambrose Sereno, *Program Executive Officer Aviation, Major Milestone Reviews: Lessons Learned Report*, Pittsburgh, Pa.: Carnegie Mellon University, Acquisition Support Program, Technical Report, CMU/SEI-2010-TR-006, ESC-TR-2010-006, September 2010. As of November 28, 2018:
https://resources.sei.cmu.edu/asset_files/TechnicalReport/2010_005_001_15212.pdf

Risen, Tom, "Disaggregation," Aerospace America, April 2017. As of October 24, 2018:
https://aerospaceamerica.aiaa.org/features/disaggregation

Rosenzweig, Phil, "Robert S. McNamara and the Evolution of Modern Management," *Harvard Business Review*, December 2010. As of November 28, 2018:
https://hbr.org/2010/12/robert-s-mcnamara-and-the-evolution-of-modern-management

SAF/AQ—See Assistant Secretary of the Air Force for Acquisition, Technology, and Logistics.

Schank, John F., Cesse Cameron Ip, Frank W. Lacroix, Robert Murphy, Mark V. Arena, Kristy N. Kamarck, and Gordon T. Lee, *Learning from Experience*, Vol. II: *Lessons from the U.S. Navy's* Ohio, Seawolf, *and* Virginia *Submarine Programs*, Santa Monica, Calif.: RAND Corporation, MG-1128/2-NAVY, 2011. As of July 27, 2018:
https://www.rand.org/pubs/monographs/MG1128z2.html

Schank, John F., Jessie Riposo, John Birkler, and James Chiesa, *The United Kingdom's Nuclear Submarine Industrial Base*, Vol. 1: *Sustaining Design and Production Resources*, Santa Monica, Calif.: RAND Corporation, MG-326/1-MOD, 2005. As of July 27, 2018:
https://www.rand.org/pubs/monographs/MG326z1.html

Schank, John F., Giles K. Smith, John Birkler, Brien Alkire, Michael Boito, Gordon T. Lee, Raj Raman, and John Ablard, *Acquisition and Competition Strategy Options for the DD(X): The U.S. Navy's 21st Century Destroyer*, Santa Monica, Calif.: RAND Corporation, MG-259/1-NAVY, 2006. As of May 25, 2018:
https://www.rand.org/pubs/monographs/MG259z1.html

Schoonover, Joanne S., *Accelerated Air Force Acquisition Processes: Lessons Learned from Desert Storm*, Maxwell Air Force Base, Montgomery, Ala: Air University Press, AU-ARf-

92-11, August 1994: As of June 6, 2019:
https://apps.dtic.mil/dtic/tr/fulltext/u2/a285289.pdf

Section 804—*See* Public Law 114-92.

Section 806—*See* Public Law 114-328.

Space Based Infrared Systems Program Office, *Space Based Infrared Systems (SBIRS) High Component Single Acquisition Management Plan*, El Segundo, Calif.: U.S. Air Force, June 30, 2002. As of June 17, 2019:
https://nsarchive2.gwu.edu/NSAEBB/NSAEBB235/new48.pdf

Spear, Tony, *NASA FBC Task Final Report*, Faster, Better, Cheaper Task Force, Washington, D.C.: National Aeronautics and Space Administration, March 13, 2000. As of November 12, 2018:
https://mars.jpl.nasa.gov/msp98/misc/fbctask.pdf

Tirpak, John A., "B-21 Bomber Critical Design Review by End of Year," *Air Force Magazine*, June 25, 2018. As of December 4, 2018:
http://www.airforcemag.com/Features/Pages/2018/June%202018/B-21-Bomber-Critical-Design-Review-by-End-of-Year.aspx

Under Secretary of Defense (Comptroller), *Department of Defense Fiscal Year (FY) 2020 Budget Estimates, Office of the Secretary of Defense, Defense-Wide Justification Book*, Vol. 3 of 5: *Research, Development, Test & Evaluation, Defense-Wide*, Washington, D.C., U.S. Department of Defense, March 2019. As of December 9, 2019:
https://comptroller.defense.gov/Portals/45/Documents/defbudget/fy2020/budget_justification/pdfs/03_RDT_and_E/RDTE_Vol3_OSD_RDTE_PB20_Justification_Book.pdf

Under Secretary of Defense for Acquisition and Sustainment, "Joint Rapid Acquisition Cell," webpage, undated. As of December 10, 2019:
https://www.acq.osd.mil/jrac/index.html

———, "Middle Tier of Acquisition (Rapid Prototyping/Rapid Fielding) Interim Governance," Washington, D.C.: U.S. Department of Defense, October 9, 2018a.

———, *Other Transactions Guide*, version 1.0, Washington, D.C.: U.S. Department of Defense, November 2018b. As of December 10, 2019:
https://www.acq.osd.mil/dpap/cpic/cp/docs/OT_Guide_(Nov_2018)_Final.pdf

Underwood, Kimberly, "Air Force Pursues SMC 2.0 Effort," *Signal Magazine*, October 30, 2018. As of December 6, 2018:
https://www.afcea.org/content/air-force-pursues-smc-20-effort

USAF—*See* U.S. Air Force.

U.S. Air Force, "Rapid Capabilities Office," fact sheet, August 28, 2009. As of June 5, 2019:
https://www.af.mil/About-Us/Fact-Sheets/Display/Article/104513/rapid-capabilities-office/

U.S. Code, Title 10, Armed Forces. As of October 17, 2019:
https://www.law.cornell.edu/uscode/text/10

U.S. Code, Title 15, Commerce and Trade. As of October 17, 2019:
https://www.law.cornell.edu/uscode/text/15

U.S. Code, Title 41, Public Contracts. As of October 17, 2019:
https://www.law.cornell.edu/uscode/text/41

USD(A&S)—*See* Under Secretary of Defense for Acquisition and Sustainment.

U.S. General Services Administration, *Procurement Through Commercial E-Commerce Portals: Implementation Plan*, Washington, D.C., March 2018. As of November 26, 2018:
https://interact.gsa.gov/sites/default/files/Commercial%20Platform%20Implementation%20Plan.pdf

———, "GSA Schedules," webpage, April 15, 2019. As of June 22, 2019:
https://www.gsa.gov/buying-selling/purchasing-programs/gsa-schedules

U.S. Government Accountability Office, *Urgent Warfighter Needs: Opportunities Exist to Expedite Development and Fielding of Joint Capabilities*, Washington, D.C., GAO-12-385, April 24, 2012a. As of December 15, 2019:
https://www.gao.gov/products/GAO-12-385

———, *Software Development: Effective Practices and Federal Challenges in Applying Agile Methods*, Washington, D.C., GAO-12-681, July 2012b. As of December 15, 2019:
https://www.gao.gov/products/GAO-12-681

———, *Schedule Assessment Guide: Best Practices for Project Schedules*, Washington, D.C., GAO-16-89G, December 2015. As of June 7, 2019:
https://www.gao.gov/products/GAO-16-89G

———, Decision B-416061, May 31, 2018a. As of December 11, 2018:
https://www.gao.gov/assets/700/692327.pdf

———, *Defense Contracting: Use by the Department of Defense of Indefinite-Delivery Contracts from Fiscal Years 2015 through 2017*, Washington, D.C., GAO-18-412R, May 10, 2018b. As of October 24, 2018:
https://www.gao.gov/products/GAO-18-412R

U.S. Navy, "Prepared to Defend," USS *Zumwalt* program, *All Hands Magazine*, Defense Media Activity, U.S. Navy Office of Information, undated.

U.S. Special Operations Command, Special Operations Forces (Acquisition, Technology, and Logistics), "Contracting," webpage, undated. As of December 11, 2018: https://www.socom.mil/SOF-ATL/Pages/contracting.aspx

Van Atta, Richard H., R. Royce Kneece, Jr., and Michael J. Lippitz, *Assessment of Accelerated Acquisition of Defense Programs*, Alexandria, Va.: Institute for Defense Analyses, Paper P-8161, September 2016. As of June 6, 2019: https://apps.dtic.mil/dtic/tr/fulltext/u2/1028525.pdf

Walden, Randall, "Air Force Rapid Capabilities Office Overview," Washington, D.C.: U.S. Air Force, June 16, 2016.

Ward, Dan, "Faster, Better, Cheaper Revisited: Program Management Lessons from NASA," *Defense AT&L*, Vol. 39, No. 2, March–April 2010, pp. 48–52. As of November 12, 2018: http://www.dtic.mil/dtic/tr/fulltext/u2/1016355.pdf

Werner, Debra, "Spotlight: Boeing Phantom Works," *Space News,* October 23, 2012. As of June 4, 2019: https://spacenews.com/spotlight-boeing-phantom-works/

———, "NRO Aims to Move Fast by Relying on Commercial Products, Expanding Internal Research and Development," *Space News*, April 17, 2018. As of November 13, 2018: https://spacenews.com/nro-aims-to-move-fast-by-relying-on-commercial-products-expanding-internal-research-and-development/

Williams, Shara, Jeffrey A. Drezner, Megan McKernan, Douglas Shontz, and Jerry M. Sollinger, *Rapid Acquisition of Army Command and Control Systems*, Santa Monica, Calif.: RAND Corporation, RR-274-A, 2014. As of June 6, 2019: https://www.rand.org/pubs/research_reports/RR274.html

Wōden, "Existing DoD OTA Consortia," webpage, September 11, 2017. As of December 11, 2018: https://www.woden.one/woden-blog/2017/9/11/existing-dod-ota-consortia

Zultner, R., "The Deming Approach to Quality Software Engineering," *Quality Progress*, Vol. 21, No. 11, 1988, pp. 58–64.